I0486260

CHAPTER ONE: NUMBERS

1. Mogaka and Onduso together can do a piece of work in 6 days. Mogaka, working alone, takes 5 days longer than Onduso. How many days does it take Onduso to do the same work alone?

2. (a) Evaluate

$$\frac{-8 \div 2 + 12 \times 9 - 4 \times 6}{56 \div 7 \times 2}$$

(b) Simplify the expression

5a − 4b − 2(a-(`2b+c)

3. Evaluate

$$\frac{28-(-18)}{-2} - \frac{15 -(-2)(-6)}{3}$$

4. Three people Odawa, Mliwa and Amina contributed money to purchase a flour mill. Odawa contributed $\frac{1}{3}$ of the total amount, Mliwa contributed $\frac{3}{8}$ of the remaining amount and Amina contributed the rest of the money. The difference in contribution between Mliwa and Amina was Kshs 40000. Calculate the price of the flour mill.

5. Evaluate:

$$\frac{-12 \div (-3) \times 4 - (-20)}{-6 \times 6 \div 3 + (-6)}$$

6. Without using logarithm tables or a calculator evaluate.

$$\sqrt{\frac{384.16 \times 0.0625}{96.04}}$$

7. Evaluate without using mathematical table

$$1000\left(\sqrt{\frac{0.0128}{200}}\right)$$

8. Express the numbers 1470 and 7056, each as a product of its prime factors.

Hence evaluate: $\dfrac{1470^2}{\sqrt{7056}}$

Leaving the answer in prime factor form

9. Evaluate:

$$\frac{\frac{3}{4} + 1\,\frac{5}{7} \div \frac{4}{7} \text{ of } 2\,\frac{1}{3}}{(1\,\frac{3}{7} - \frac{5}{8}) \times \frac{2}{3}}$$

10. Pipes A can fill an empty water tank in 3 hours while pipe B can fill the same tank in 6 hours. When the tank is full it can be emptied by pipe C in 8 hours. Pipe A and B are opened at the same time when the tank is empty. If one hour later, pipe C is also opened, find the total time taken to fill the tank.

11. In a fund- raising committee of 45 people the ratio of men to women 7.2. Find the number of women required to join the existing committee so that the ratio of men to women is changes to 5:4

12. Without using mathematical tables or calculators, evaluate

$$\sqrt[3]{\dfrac{675 \times 135}{\sqrt{2025}}}$$

13. All prime numbers less than ten are arranged in descending order to form a number

    (a) Write down the number formed

    (b) What is the total value of the second digit?

14. Evaluate without using mathematical tables or a calculator $\dfrac{0.0084 \times 1.23 \times 3.5,}{2.87 \times 0.056}$

    Expressing the answer as a fraction in its simplest form.

15. Evaluate $\frac{1}{3}$ of $(2 \frac{3}{4} - 5 \frac{1}{2}) \times 3\frac{6}{7} \div \frac{9}{4}$

16. Evaluate without using mathematical tables or the calculator

$$\sqrt{\dfrac{(.0.0625 \times 2.56)}{0.25 \times 0.08 \times 0.5}}$$

17. Evaluate without using mathematical tables or the calculator

$$\dfrac{1.9 \times 0.032}{20 \times 0.0038}$$

18. Evaluate $\dfrac{2 \frac{3}{4} \times \frac{8}{33}}{3 + (5\frac{2}{5} \div \frac{9}{25})}$

19. Without using tables or calculators evaluate

$$\sqrt{\dfrac{153 \times 0.18}{0.68 \times 0.32}}$$

20. Without using mathematical tables, evaluate

$$\sqrt{1.2 \times \dfrac{0.0324}{0.0072}}$$

21. Simplify $^2/_3$ of $12 - (1\,^1/_3 + 1\,^1/_4)$

22. If x= 2, Find the value of $x^3 - 5x^2 - 4x + 3$

23. If X = ½ y= ¼ and z = $^2/_3$ Find the value of

$$\dfrac{x + yz}{y - xz}$$

24. Find a and b if $3.168 = 3^a/_b$

25. Find the greatest common factor of $x^8 y^2$ and $4xy^4$. Hence factorize completely the expression $x^3y^2 - 4xy^4$

26. A hot water tap can fill a bath in 5 minutes while a cold water tap can fill the same bath in 3 minutes. The drain pipe can empty the full bath in 3 ¾ minutes. The two taps and the drain pipe are fully open for 1 ½ minutes after which the drain pipe are fully open for 1 ½ minutes after which the drain pipe is closed. How much will take it take to fill the bath?

27. A farmer distributed his cabbages as follows

A certain hospital received a quarter of the total number of bags. A nearby school received half of the remainder. A green grocer received a third of what the school received. What remained were six bags more than what the green grocer received. How many bags of cabbages did the farmer have?

CHAPTER TWO: ALGEBRAIC EXPRESSIONS

1. Given that $y = \dfrac{2x - z}{x + 3z}$ express x in terms of y and z

2. Simplify the expression

   $$\dfrac{x - 1}{x} - \dfrac{2x + 1}{3x}$$

   Hence solve the equation

   $$\dfrac{x - 1}{x} - \dfrac{2x + 1}{3x} = \dfrac{2}{3}$$

3. Factorize $a^2 - b^2$

   Hence find the exact value of $2557^2 - 2547^2$

4. Simplify $\dfrac{p^2 - 2pq + q^2}{P^3 - pq^2 + p^2 q - q^3}$

5. Given that $y = 2x - z$, express x in terms of y and z.

   Four farmers took their goats to a market. Mohammed had two more goats as Koech had 3 times as many goats as Mohammed, whereas Odupoy had 10 goats less than both Mohammed and Koech.

   (i) Write a simplified algebraic expression with one variable, representing the total number of goats.

   (ii) Three butchers bought all the goats and shared them equally. If each butcher got 17 goats, how many did odupoy sell to the butchers?

6. Factorize completely $3x^2 - 2xy - y^2$

7. Solve the equation

$$\frac{1}{4x} = \frac{5}{6x} - 7$$

8. Simplify

$$\frac{a}{2(a+b)} + \frac{b}{2(a-b)}$$

9. Factorize completely $28x^2 + 3x - 1$

10. Three years ago, Juma was three times as old. As Ali in two years time, the sum of their ages will be 62. Determine their ages

11. Two pairs of trousers and three shirts cost a total of Kshs. Five such pairs of trousers and two shirt cost a total of Kshs 810. Find the price of a pair of trouser and shirt.

CHAPTER THREE: RATES, RATIO PERCENTAGE AND PROPORTION

1. Akinyi bought and beans from a wholesaler. She then mixed the maize and beans in the ratio 4:3 she bought the maize at Kshs 21 per kg and the beans 42 per kg. If she was to make a profit of 30%. What should be the selling price of 1 kg of the mixture?

2. Water flows from a tap at the rate of 27 cm^3 per second into a rectangular container of length 60 cm, breadth 30 cm and height 40 cm. If at 6.00 PM the container was half full, what will be the height of water at 6.04 pm?

3. Two businessmen jointly bought a minibus which could ferry 25 paying passengers when full. The fare between two towns A and B was Kshs 80 per passenger for one way. The minibus made three round trips between the two towns daily. The cost of fuel was Kshs 1500 per day. The driver and the conductor were paid daily allowances of Kshs 200 and Kshs 150 respectively.

A further Kshs 4000 per day was set aside for maintenance, insurance and loan repayment.

    (a)   (i)   How much money was collected from the passengers that day?

           (ii)   How much was the net profit?

    (b)   On another day, the minibus was 80% full on the average for the three round trips, how much did each businessman get if the day's profit was shared in the ratio 2:3?

4. Wainaina has two dairy farms, A and B. Farm A produces milk with 3 ¼ percent fat and farm B produces milk with 4 ¼ percent fat.

    (a) Determine

        (i) The total mass of milk fat in 50 kg of milk from farm A and 30 kg of milk from farm B

        (ii) The percentage of fat in a mixture of 50kg of milk A and 30kg of milk from B

    (b) The range of values of mass of milk from farm B that must be used in a 50kg mixture so that the mixture may have at least 4 percent fat.

5. In the year 2001, the price of a sofa set in a shop was Kshs 12,000

    (a) Calculate the amount of money received from the sales of 240 sofa sets that year.

    (b) (i) In the year 2002 the price of each sofa set increased by 25% while the number of sets sold decreased by 10%. Calculate the percentage increase in the amount received from the sales

  (ii) If the end of year 2002, the price of each sofa set changed in the ratio 16: 15, calculate the price of each sofa set in the year 2003.

    (c) The number of sofa sets sold in the year 2003 was P% less than the number sold in the year 2001.

Calculate the value of P, given that the amounts received from sales if the two years were equal.

6. A solution whose volume is 80 litres is made up of 40% of water and 60% of alcohol. When x litres of water is added, the percentage of alcohol drops to 40%.

    (a)   Find the value of x

    (b)   Thirty litres of water is added to the new solution. Calculate the percentage of alcohol in the resulting solution

    (c)   If 5 litres of the solution in (b) above is added to 2 litres of the original solution, calculate in the simplest form, the ratio of water to that of alcohol in the resulting solution.

7. Three business partners, Asha, Nangila and Cherop contributed Kshs 60,000, Kshs 85,000 and Kshs 105, 000 respectively. They agreed to put 25% of the profit back into business each year. They also agreed to put aside 40% of the remaining profit to cater for taxes and insurance. The rest of the profit would then be shared among the partners in the ratio of their contributions. At the end of the first year, the business realized a gross profit of Kshs 225, 000.

    (a)   Calculate the amount of money Cherop received more than Asha at the end of the first year.

    (b)   Nangila further invested Kshs 25,000 into the business at the beginning of the second year. Given that the gross profit at the end of the second year increased in the ratio 10:9, calculate Nangila's share of the profit at the end of the second year.

8. Kipketer can cultivate a piece of land in 7 hrs while Wanjiku can do the same work in 5 hours. Find the time they would take to cultivate the piece of land when working together.

9. Mogaka and Ondiso working together can do a piece of work in 6 days. Mogaka working alone, takes 5 days longer than Onduso. How many days does it take Onduso to do the work alone.

10. A certain amount of money was shared among 3 children in the ratio 7:5:3 the largest share was Kshs 91. Find the

   (a)   Total amount of money

   (b)   Difference in the money received as the largest share and the smallest share.

CHAPTER FOUR: MEASUREMENTS

1. The figure below shows a portable kennel

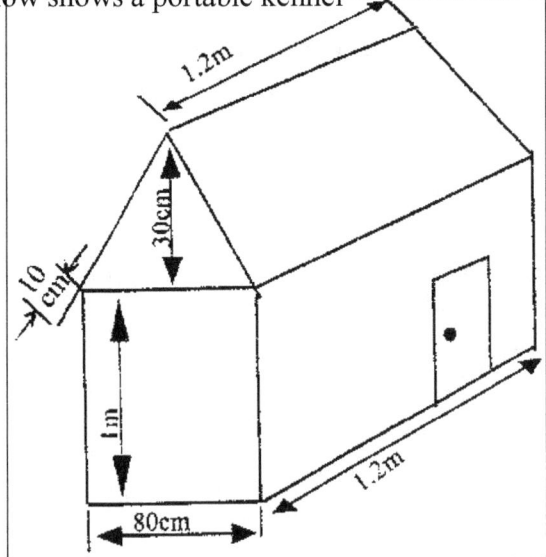

(a)   Calculate

   (i)   The total surface area of the walls and the floor (include the door as part of the wall.

   (ii)   The total surface area of the roof

(b)   The cost of roofing is Kshs 300 per square metre and that of making walls and floor Kshs 350 per square metre. Find the cost of making the kennel.

2. The enclosed region shown in the figure below represents a ranch draw to scale. The actual area of the ranch is 1075 hectares

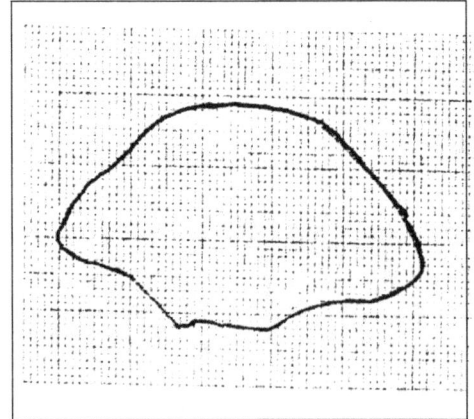

(a)   Estimate the area of the enclosed region in square centimeters

(b)   Calculate the linear scale used

3. The figure below shows an octagon obtained by cutting off four congruent triangles from a rectangle measuring 19.5 by 16.5 cm

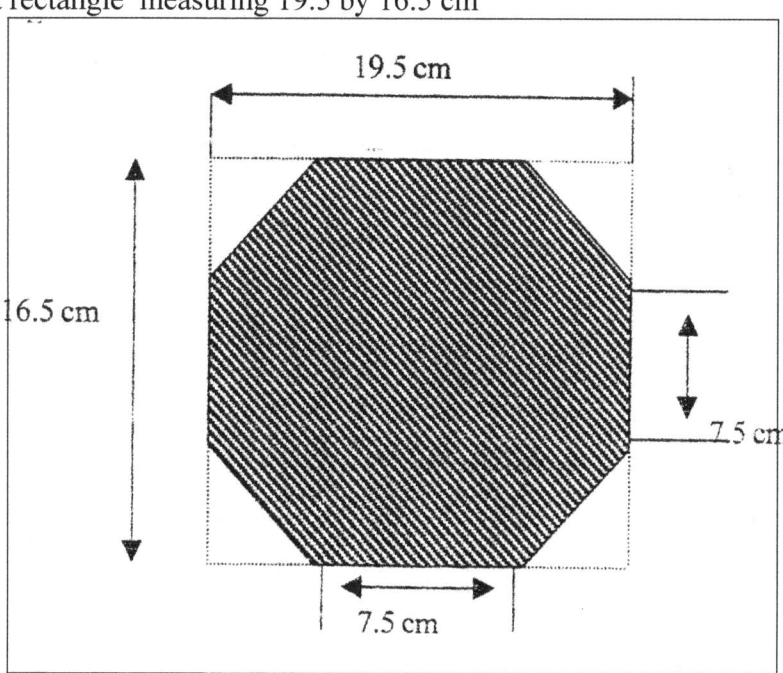

Calculate the area of the octagon

4. The length of a hollow cylindrical pipe is 6 metres. Its external diameter is 11cm and has a thickness of 1 cm. Calculate the, volume in cm³ of the materials used to make the pipe. Take π as 3.142.

5. The area of rhombus is 60 cm². Given that one of its diagonals is 15 cm long, calculate the perimeter of the rhombus.

6. A cylindrical piece of wood of radius 4.2 cm and length 150 cm is cut lengthwise into two equal pieces.

    Calculate the surface area of one piece

    (Take π as $^{22}/_7$)

7. The diagram below (not drawn to scale) represents the cross section of a solid prism of height 8.0 cm

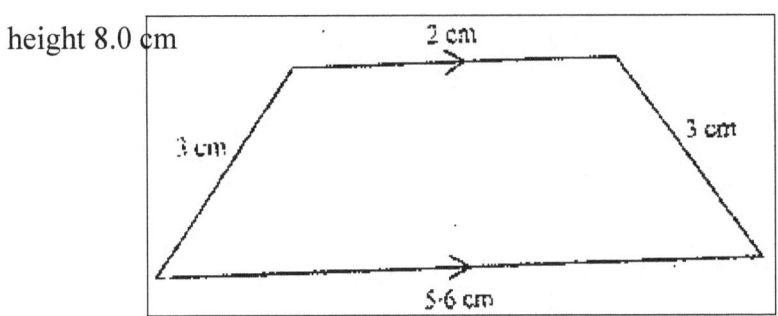

(a)   Calculate the volume of the prism

(b)   Given that the density of the prism is 5.75 g/cm³, calculate its mass in

grams

(c)   A second prism is similar to the first one but is made of different material. The

volume of the second prism is 246.24 cm³

(i)   Calculate the area of cross section of the second prism

(ii)   Given the ratio of the mass of the first prism to the second is 2:5, find the

density of the second prism.

8. A square brass plate is 2 mm thick and has a mass of 1.05 kg. The density of the brass is

8.4g/ cm³. Calculate the length of the plane in centimeters.

9. Two cylindrical containers are similar. The larger one has internal cross- section area of

45cm² and can hold 0.95 litres of liquid when full. The smaller container has internal

cross- section area of 20cm²

(a)   Calculate the capacity of the smaller container

(b)   The larger container is filled with juice to a height of 13 cm. Juice is then drawn from it and empties into the smaller container until the depth of the juice in both containers are equal. Calculate the depth of juice in each container.

(c)   One fifth of the juice in the larger container in part (b) above is further drawn and emptied into the smaller container. Find the differences in the depths of the juice in the two containers.

10. Pieces of soap are packed in a cuboid container measuring 36 cm by 24 cm by 18 cm. Each piece of soap is similar to the container. If the linear scale factor between the container and the soap is 1/6. Find the volume of each piece of soap.

11. A cylindrical water tank is of diameter 7 metres and height 2.8 metres

(a)   Find the capacity of the water tank in litres

(b)   Six members of family use15 litres each per day. Each day 80 litres are used for cooking and washing. And a further 60 litres are wasted.
        Find the number of complete days a full tank would last the family

(c)   Two members of the family were absent for 90 days. During the 90 days, wastage was reduced by 20% but cooking and washing remained the same.
        Calculate the number of days a full tank would now last the family

12. A company is to construct parking bay whose area is 135m². It is to be covered with a concrete slab of uniform thickness of 0.15m. To make the slab cement, ballast and sand

are to be mixed so that their masses are in the ratio 1:4:4 the mass of 1m³ of dry slab is 2,500 kg.

(a) Calculate

   (i) The volume of the slab

   (ii) The mass of dry slab

   (iii) The mass of cement to be used

(b) If one bag of cement is 50kg. Find the number of bags to be purchased

(c) If a lorry carries 7 tonnes of sand, calculate the number of lorries of sand to be purchased

13. An Artisan has 63 kg of metal of density 7000 kg/m³. He intends to use to make a rectangular pipe with external dimensions 12 cm by 15 cm and internal dimensions 10cm by 12 cm. Calculate the length of the pipe in metres.

14. The figure below represents hollow cylinder. The internal and external radii are estimated to be 6 cm and 8 cm respectively, to the nearest whole number. The height of the cylinder is exactly 14 cm.

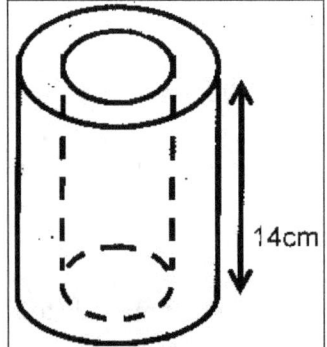

(a) Determine the exact values for internal and external radii which will give maximum volume of the material used.

(b)   Calculate the maximum possible volume of the material used. Take the value of

TT to be $^{22}/_7$

15. Calculate the volume of a prism whose length is 25 cm and whose length is 25 cm and

whose cross – section is an equilateral triangle of side 3 cm

16. The figure below shows an octagon obtained by cutting off four congruent triangles

from a rectangle measuring 19.5 by 16.5 cm

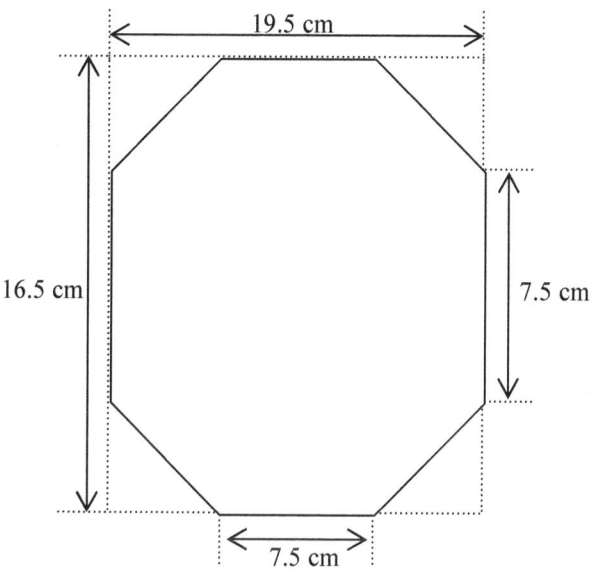

Calculate the area of the octagon

17. The figure below represents a kite ABCD, AB = AD = 15 cm. the diagonals BD and AC intersect at O, AC = 30 cm and AO = 12 cm.

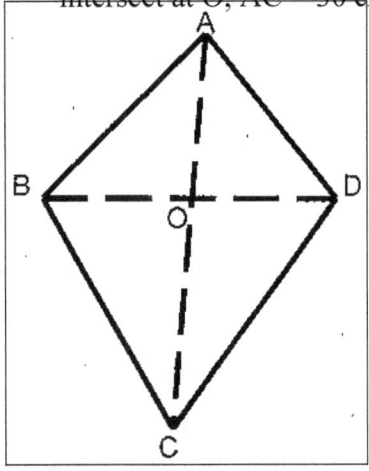

Find the area of the kite

18. The figure below is a map of a forest drawn on a grid of 1 cm squares

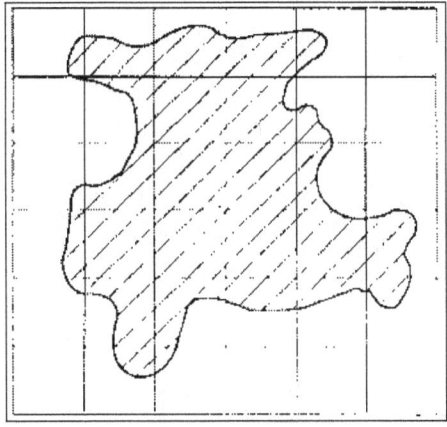

(a)   Estimate the area of the map in square centimeters if the scale of the map is 1: 50, 000; estimate the area of the forest in hectares.

CHAPTER FIVE: LINEAR EQUATIONS

1. A cloth dealer sold 3 shirts and 2 trousers for Kshs 840 and 4 shirts and 5 trousers for Kshs 1680 find the cost of 1 shirt and the cost of 1 trouser

2. Solve the simultaneous equations

   $2x - y = 3$

   $x^2 - xy = -4$

3. The cost of 5 skirts and blouses is Kshs 1750. Mueni bought three of the skirts and one of the blouses for Kshs 850. Find the cost of each item.

4. Akinyi bought three cups and four spoons for Kshs 324. Wanjiru bought five cups and Fatuma bought two spoons of the same type as those bought by Akinyi, Wanjiku paid Kshs 228 more than Fatuma. Find the price of each cup and each spoon.

5. Mary has 21 coins whose total value is Kshs. 72. There are twice as many five shillings coins as there are ten shilling coins. The rest one shillings coins. Find the number of ten shillings coins that Mary has.   (4 mks)

6. The mass of 6 similar art books and 4 similar biology books is 7.2 kg. The mass of 2 such art books and 3 such biology books is 3.4 kg. Find the mass of one art book and the mass of one biology book

7. Karani bought 4 pencils and 6 biros – pens for Kshs 66 and Tachora bought 2 pencils and 5 biro pens for Kshs 51.

(a)   Find the price of each item

(b)   Musoma spent Kshs. 228 to buy the same type of pencils and biro – pens if the number of biro pens he bought were 4 more than the number of pencils, find the number of pencils bought.

8. Solve the simultaneous equations below

   $2x - 3y = 5$

   $-x + 2y = -3$

9. The length of a room is 4 metres longer than its width. Find the length of the room if its area is $32m^2$

10. Hadija and Kagendo bought the same types of pens and exercise books from the same types of pens and exercise books from the same shop. Hadija bought 2 pens and 3 exercise books for Kshs 78. Kagendo bought 3 pens and 4 exercise books for Kshs 108.

    Calculate the cost of each item

11. In fourteen years time, a mother will be twice as old as her son. Four years ago, the sum of their ages was 30 years. Find how old the mother was, when the son was born.

12. Three years ago Juma was three times as old as Ali. In two years time the sum of their ages will be 62. Determine their ages.

13. Two pairs of trousers and three shirts costs a total of Kshs 390. Five such pairs of trousers and two shirts cost a total of Kshs 810. Find the price of a pair of trousers and a shirt.

14. A shopkeeper sells two- types of pangas type x and type y. Twelve x pangas and five type y pangas cost Kshs 1260, while nine type x pangas and fifteen type y pangas cost 1620. Mugala bought eighteen type y pangas. How much did he pay for them?

CHAPTER SIX: COMMERCIAL ARITHMETICS

1. The cash prize of a television set is Kshs 25000. A customer paid a deposit of Kshs 3750. He repaid the amount owing in 24 equal monthly installments. If he was charged simple interest at the rate of 40% p.a how much was each installment?

2. Mr Ngeny borrowed Kshs 560,000 from a bank to buy a piece of land. He was required to repay the loan with simple interest for a period of 48 months. The repayment amounted to Kshs 21,000 per month.

   Calculate

   (a) The interest paid to the bank

   (b) The rate per annum of the simple interest

3. A car dealer charges 5% commission for selling a car. He received a commission of Kshs 17,500 for selling car. How much money did the owner receive from the sale of his car?

4. A company saleslady sold goods worth Kshs 240,000 from this sale she earned a commission of Kshs 4,000

   (a) Calculate the rate of commission

   (b) If she sold good whose total marked price was Kshs 360,000 and allowed a discount of 2% calculate the amount of commission she received.

5. A business woman bought two bags of maize at the same price per bag. She discovered that one bag was of high quality and the other of low quality. On the high quality bag she made a profit by selling at Kshs 1,040, whereas on the low quality bag she made a loss by selling at Kshs 880. If the profit was three times the loss, calculate the buying price per bag.

6. A salesman gets a commission of 2. 4 % on sales up to Kshs 100,000. He gets an additional commission of 1.5% on sales above this. Calculate the commission he gets on sales worth Kshs 280,000.

7. Three people Koris, Wangare and Hassan contributed money to start a business. Korir contributed a quarter of the total amount and Wangare two fifths of the remainder. Hassan's contribution was one and a half times that of Koris. They borrowed the rest of the money from the bank which was Kshs 60,000 less than Hassan's contribution. Find the total amount required to start the business.

8. A Kenyan tourist left Germany for Kenya through Switzerland. While in Switzerland he bought a watch worth 52 deutsche Marks. Find the value of the watch in:

    (a)   Swiss Francs.

    (b)   Kenya Shillings

            Use the exchange rtes below:

            1 Swiss Franc = 1.28 Deutsche Marks.

            1 Swiss Franc = 45.21 Kenya Shillings

9.   A salesman earns a basic salary of Kshs. 9000 per month

        In addition he is also paid a commission of 5% for sales above Kshs 15000

        In a certain month he sold goods worth Kshs. 120, 000 at a discount of 2½ %.

        Calculate his total earnings that month

10. In this question, mathematical table should not be used

A Kenyan bank buys and sells foreign currencies as shown below

	Buying	Selling
	(In Kenya shillings)	In Kenya Shillings
1 Hong Kong dollar	9.74	9.77
1 South African rand	12.03	12.11

A tourists arrived in Kenya with 105 000 Hong Kong dollars and changed the whole amount to Kenyan shillings. While in Kenya, she pent Kshs 403 897 and changed the balance to South African rand before leaving for South Africa. Calculate the amount, in South African rand that she received.

11. A Kenyan businessman bought goods from Japan worth 2, 950 000 Japanese yen. On arrival in Kenya custom duty of 20% was charged on the value of the goods.

    If the exchange rates were as follows

    1 US dollar = 118 Japanese Yen

    1 US dollar = 76 Kenya shillings

    Calculate the duty paid in Kenya shillings

12. Two businessmen jointly bought a minibus which could ferry 25 paying passengers when full. The fare between two towns A and B was Kshs. 80 per passenger for one way. The minibus made three round trips between the two towns daily. The cost of fuel was Kshs 1500 per day. The driver and the conductor were paid daily allowances of Kshs 200 and Kshs 150 respectively.

A further Kshs 4000 per day was set aside for maintenance.

(a) One day the minibus was full on every trip.

    (i)   How much money was collected from the passengers that day?

    (ii)   How much was the net profit?

(b) On another day, the minibus was 80% on the average for the three round trips. How much did each business get if the days profit was shared in the ratio 2:3?

13. A traveler had sterling pounds 918 with which he bought Kenya shillings at the rate of Kshs 84 per sterling pound. He did not spend the money as intended. Later, he used the Kenyan shillings to buy sterling pound at the rate of Kshs. 85 per sterling pound. Calculate the amount of money in sterling pounds lost in the whole transaction.

14. A commercial bank buys and sells Japanese Yen in Kenya shillings at the rates shown below

Buying   0.5024

Selling   0.5446

A Japanese tourist at the end of his tour of Kenya was left with Kshs. 30000 which he converted to Japanese Yen through the commercial bank. How many Japanese Yen did he get?

15. In the month of January, an insurance salesman earned Kshs. 6750 which was commission of 4.5% of the premiums paid to the company.

(a) Calculate the premium paid to the company.

(b) In February the rate of commission was reduced by $66^2/_3$% and the premiums reduced by 10% calculate the amount earned by the salesman in the month of February

16. Akinyi, Bundi, Cura and Diba invested some money in a business in the ratio of 7:9:10:14 respectively. The business realized a profit of Kshs 46800. They shared 12% of the profit equally and the remainder in the ratio of their contributions. Calculate the total amount of money received by Diba.

17. A telephone bill includes Kshs 4320 for a local calls Kshs 3260 for trank calls and rental charge Kshs 2080. A value added tax (V.A.T) is then charged at 15%, Find the total bill.

18. During a certain period. The exchange rates were as follows

1 sterling pound = Kshs 102.0

1 sterling pound = 1.7 us dollar

1 U.S dollar = Kshs 60.6

A school management intended to import textbooks worth Kshs 500,000 from UK. It changed the money to sterling pounds. Later the management found out that the books the sterling pounds to dollars. Unfortunately a financial crisis arose and the money had to be converted to Kenya shillings. Calculate the total amount of money the management ended up with.

19. A fruiterer bought 144 pineapples at Kshs 100 for every six pineapples. She sold some of them at Kshs 72 for every three and the rest at Kshs 60 for every two. If she made a 65% profit, calculate the number of pineapples sold at Kshs 72 for every three.

CHAPTER SEVEN: GEOMETRY

1. A point B is on a bearing of 080^0 from a port A and at a distance of 95 km. A submarine is stationed at a port D, which is on a bearing of 200^0 from AM and a distance of 124 km from B.

   A ship leaves B and moves directly southwards to an island P, which is on a bearing of 140 from A. The submarine at D on realizing that the ship was heading fro the island P, decides to head straight for the island to intercept the ship

   Using a scale 0f 1 cm to represent 10 km, make a scale drawing showing the relative positions of A, B, D, P.

   Hence find

   (i) The distance from A to D

   (ii) The bearing of the submarine from the ship was setting off from B

   (iii)The bearing of the island P from D

   (iv)The distance the submarine had to cover to reach the island P

2. Four towns R, T, K and G are such that T is 84 km directly to the north R, and K is on a bearing of 295^0 from R at a distance of 60 km. G is on a bearing of 340^0 from K and a distance of 30 km. Using a scale of 1 cm to represent 10 km, make an accurate scale drawing to show the relative positions of the town.

   Find

   (a) The distance and the bearing of T from K

   (b) The distance and the bearing G from T

   (c) The bearing of R from G

3. Two aeroplanes, S and T leave airports A at the same time. S flies on a bearing of 060 at 750 km/h while T flies on a bearing of 210^0 at 900km/h.

    (a) Using a suitable scale, draw a diagram to show the positions of the aeroplane after two hours.

    (b) Use your diagram to determine

         (i) The actual distance between the two aeroplanes

         (ii) The bearing of T from S

         (iii) The bearing of S from T

4. A point A is directly below a window. Another point B is 15 m from A and at the same horizontal level. From B angle of elevation of the top of the bottom of the window is 300 and the angle of elevation of the top of the window is 350. Calculate the vertical distance.

    (a) From A to the bottom of the window

    (b) From the bottom to top of the window

5. Find by calculation the sum of all the interior angles in the figure ABCDEFGHI below

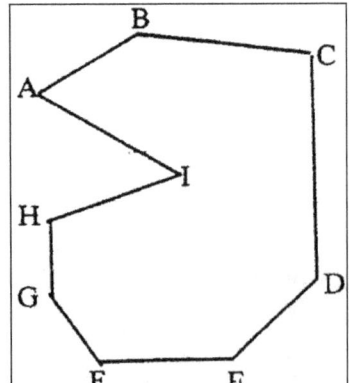

6. Shopping centers X, Y and Z are such that Y is 12 km south of X and Z is 15 km from X. Z is on a bearing of 330^0 from Y. Find the bearing of Z from X.

28

7.	An electric pylon is 30m high. A point S on the top of the pylon is vertically above another point R on the ground. Points A and B are on the same horizontal ground as R. Point A due south of the pylon and the angle of elevation of S from A is 26⁰. Point B is due west of the pylon and the angle of elevation of S from B is 32⁰

Find the

(a)	Distance from A and B

(b)	Bearing of B from A

8.	The figure below is a polygon in which AB = CD = FA = 12cm BC = EF = 4cm and BAF =- CDE = 120⁰. AD is a line of symmetry.

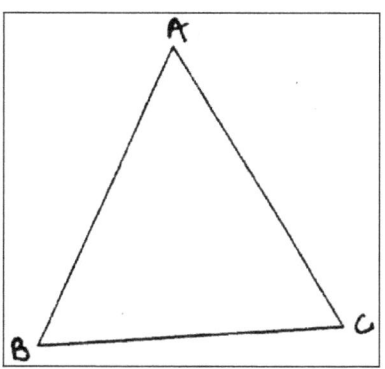

Find the area of the polygon.

9.	The figure below shows a triangle ABC.

a)   Using a ruler and a pair of compasses, determine a point D on the line BC

b)   such that BD:DC = 1:2.

b)   Find the area of triangle ABD, given that AB = AC.

10.   A boat at point x is 200 m to the south of point Y. The boat sails X to another

point Z. Point Z is 200m on a bearing of 310^0 from X, Y and Z are on the same

horizontal plane.

(a) Calculate the bearing and the distance of Z from Y

(b) W is the point on the path of the boat nearest to Y.

   Calculate the distance WY

(c) A vertical tower stands at point Y. The angle of point X from the top of the tower is

   6^0 calculate the angle of elevation of the top of the tower from W.

11.   The figure below shows a quadrilateral ABCD in which AB = 8 cm, DC = 12 cm, $\angle$

BAD = 45^0, $\angle$ CBD = 90^0 and BCD = 30^0.

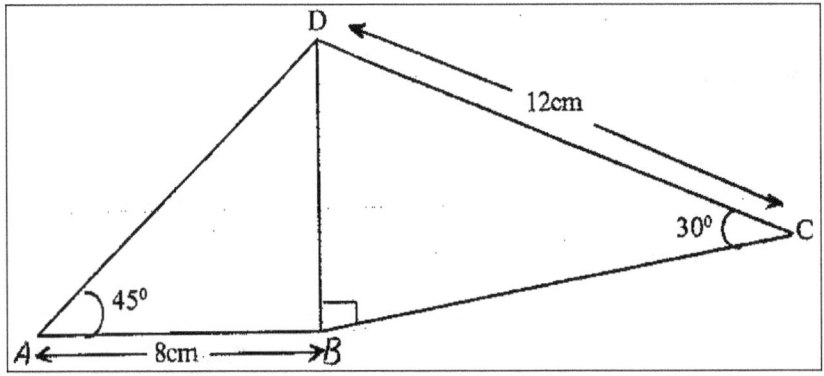

Find:

   (a) The length of BD

(b) The size of the angle A D B

12.   In the figure below, ABCDE is a regular pentagon and ABF is an equilateral triangle

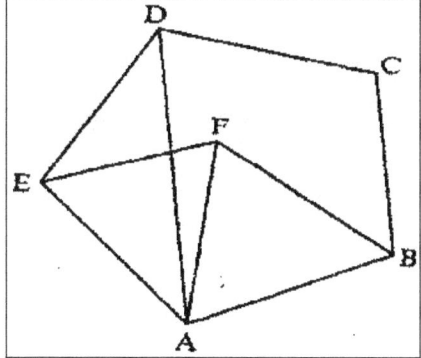

Find the size of

a) ∠ ADE

b) ∠ AEF

c) ∠ DAF

13.   In this question use a pair of compasses and a ruler only

(a) construct triangle ABC such that AB = 6 cm, BC = 8cm and ∠ABC 135^0

(b) Construct the height of triangle ABC in a) above taking BC as the base

14.   The size of an interior angle of a regular polygon is $3x^0$ while its exterior angle is $(x-20)^0$. Find the number of sides of the polygon

15.   Points L and M are equidistant from another point K. The bearing of L from K is 330^0. The bearing of M from K is 220^0.

Calculate the bearing of M from L

16. Four points B,C,Q and D lie on the same plane point B is the 42 km due south- west of town Q. Point C is 50 km on a bearing of 560^0 from Q. Point D is equidistant from B, Q and C.

   (a) Using the scale 1 cm represents 10 km, construct a diagram showing the position of B, C, Q and D

   (b) Determine the

      (i) Distance between B and C

      (ii) Bearing D from B

17. Two aeroplanes P and Q, leave an airport at the same time flies on a bearing of 240^0 at 900km/hr while Q flies due East at 750 km/hr

   (a) Using a scale of 1v cm drawing to show the positions of the aeroplanes after 40 minutes.

   (b) Use the scale drawing to find the distance between the two aeroplane after 40 minutes

   (c) Determine the bearing of

      (i) P from Q ans 254^0

      (ii) Q from P ans 74^0

18. A port B is no a bearing of 080 from a port A and at a distance of 95 km. A submarine is stationed port D which is on a bearing of 200^0 from A, and a distance of 124 km from B.

A ship leaves B and moves directly southwards to an island P, which is on a bearing of 140^0 from A. The submarine at D on realizing that the ship was heading for the island P decides to head straight for the island to intercept the ship.

Using a scale of 1 cm to represent 10 km, make a scale drawing showing the relative position of A, B D and P.

Hence find:

(i)   The distance from A and D

(ii)   The bearing of the submarine from the ship when the ship was setting off from B

(iii)   The baring of the island P from D

(iv)   The distance the submarine had to cover to reach the island

19.   Four towns R, T, K and G are such that T is 84 km directly to the north R and K is on a bearing of 295^0 from R at a distance of 60 km. G is on a bearing of 340^0 from K and a distance of 30 km. Using a scale of 1 cm to represent 10 km, make an acute scale drawing to show the relative positions of the towns.

Find

(a) The distance and bearing of T from K

(b) The bearing of R from G

20. In the figure below, ABCDE is a regular pentagon and M is the midpoint of AB. DM intersects EB at N. (T7)

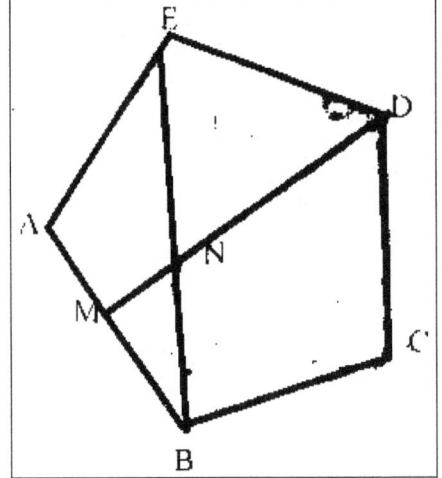

Find the size of

(a) ∠ BAE

(b) ∠ BED

(c) ∠ BNM

21. Use a ruler and compasses in this question. Draw a parallelogram ABCD in which AB = 8cm, BC = 6 cm and BAD = 75. By construction, determine the perpendicular distance between AB and CD.

22. The interior angles of the hexagon are $2x^0$, $\frac{1}{2} x^0$, $x + 40^0$, 110^0, 130^0 and 160^0. Find the value of the smallest angle.

23. The size of an interior angle of a regular polygon is 156^0. Find the number of sides of the polygon.

CHAPTER EIGHT: COMMON SOLIDS

1. The figure below shows a net of a prism whose cross – section is an equilateral triangle.

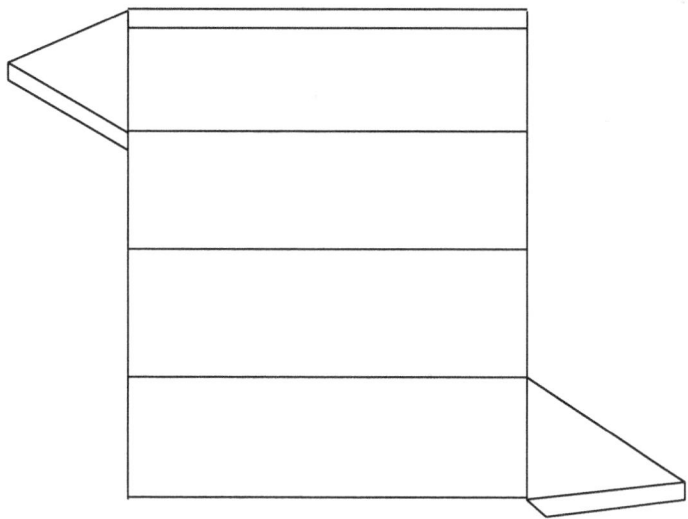

   a) Sketch the prism

   b) State the number of planes of symmetry of the prism.

2. The figure below represents a square based solid with a path marked on it.

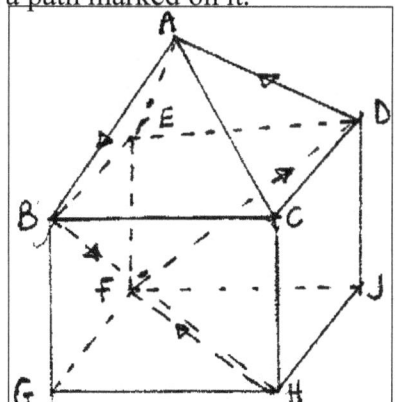

Sketch and label the net of the solid.

3. The figure below represents below represents a prism of length 7 cm

AB = AE = CD = 2 cm and BC – ED = 1 cm

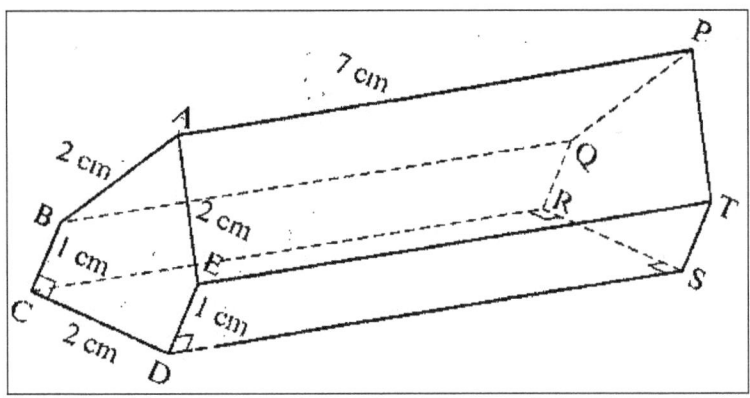

Draw the net of the prism

4. The diagram below represents a right pyramid on a square base of side 3 cm. The slant of the

   pyramid is 4 cm.

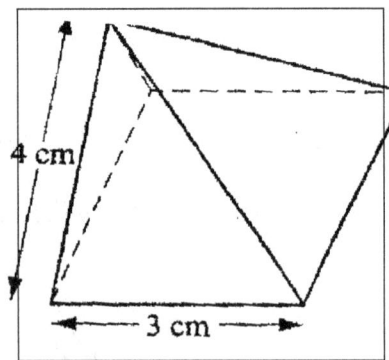

   (a)   Draw a net of the pyramid

   (b)   On the net drawn, measure the height of a triangular face from the top of

           the Pyramid

5.   (a)   Draw a regular pentagon of side 4 cm

      (b)   On the diagram drawn, construct a circle which touches all the sides of the

              pentagon

6. The figure below shows a solid regular tetrapack of sides 6 cm

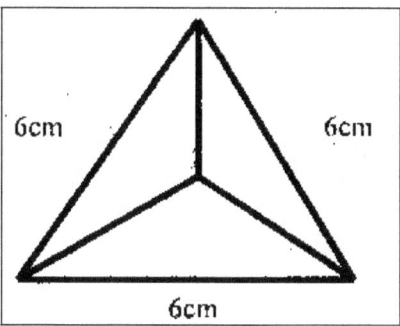

6cm   6cm

6cm

   (a)   Draw a net of the solid

   (b)   Find the surface area of the solid

7. The figure below shows a solid made by pasting two equal regular tetrahedral

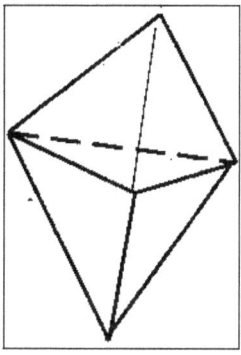

   (a)   Draw a net of the solid

   (b)   If each face is an equilateral triangle of side 5cm, find the surface area of the

          solid.

8. (a) Sketch the net of the prism shown below

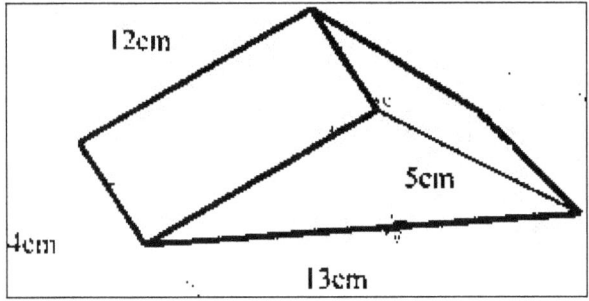

(b) Find the surface area of the solid

CHAPTER NINE: NUMBERS II

1. Use logarithms to evaluate

$$\sqrt[3]{\frac{36.15 \times 0.02573}{1,938}}$$

2. Find the value of x which satisfies the equation.

   $16^{x2} = 8^{4x-3}$

3. Use logarithms to evaluate $\dfrac{(1934)^2 \times \sqrt{0.00324}}{436}$

4. Use logarithms to evaluate

   $55.9 \div (02621 \times 0.01177)^{1/5}$

5. Simplify $2^x \times 5^{2x} \div 2^{-x}$

6. Use logarithms to evaluate

   $(3.256 \times 0.0536)^{1/3}$

7. Solve for x in the equation

   $32^{(x-3)} \div 8^{(x-4)} = 64 \div 2^x$

8. Solve for x in the equations $\dfrac{81^{2x} \times 27^x = 729}{9x}$

9. Use reciprocal and square tables to evaluate to 4 significant figures, the expression:

   $\left(\dfrac{1}{24.56} \right) + 4.346^2$

10. Use logarithm tables, to evaluate

$$\left(\frac{0.032 \times 14.26}{0.006}\right)^{2/3}$$

11. Find the value of x in the following equation

$$49^{(x+1)} + 7^{(2x)} = 350$$

12. Use logarithms to evaluate

$$\frac{(0.07284)^2}{3\sqrt{0.06195}}$$

13. Find the value of m in the following equation

$$(1/27^m \times (81)^{-1} = 243$$

14. Given that $P = 3^y$ express the equation $3^{(2y-1)} + 2 \times 3^{(y-1)} = 1$ in terms of P hence or otherwise

find the value of y in the equation $3^{(2y-1)} + 2 \times 3^{(y-1)} = 1$

15. Use logarithms to evaluate $55.9 \div (0.2621 \times 0.01177)^{1/5}$

16. Use logarithms to evaluate

$$\left(\frac{6.79 \times 0.3911}{\text{Log } 5}\right)^{3/4}$$

17. Use logarithms to evaluate

$$\sqrt[3]{\dfrac{1.23 \times 0.0089}{79.54}}$$

18. Solve for x in the equation

$$X = \dfrac{0.0056^{\frac{1}{2}}}{1.38 \times 27.42}$$

CHAPTER TEN: EQUATIONS OF LINES

1. The coordinates of the points P and Q are (1, -2) and (4, 10) respectively.

    A point T divides the line PQ in the ratio 2: 1

    (a) Determine the coordinates of T

    (b) (i)   Find the gradient of a line perpendicular to PQ

        (ii)   Hence determine the equation of the line perpendicular PQ and passing through T

        (iii)   If the line meets the y- axis at R, calculate the distance TR, to three significant

              figures

2. A line L_1 passes though point (1, 2) and has a gradient of 5. Another line L_2, is perpendicular

    to L_1 and meets it at a point where x = 4. Find the equation for L_2 in the form of y = mx + c

3. P (5, -4) and Q (-1, 2) are points on a straight line. Find the equation of the perpendicular

    bisector of PQ: giving the answer in the form y = mx+c.

4. On the diagram below, the line whose equation is 7y – 3x + 30 = 0 passes though the

    points A and B. Point A on the x-axis while point B is equidistant from x and y axes.

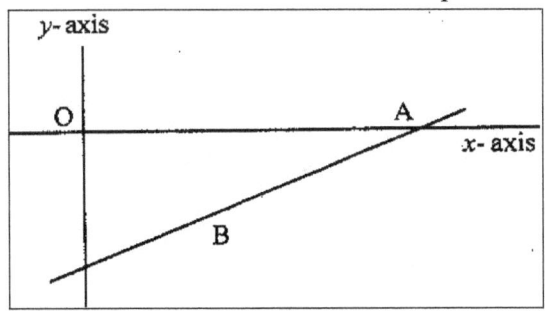

Calculate the co-ordinates of the points A and B

5. A line with gradient of -3 passes through the points (3. k) and (k.8). Find the value of k and

    hence express the equation of the line in the form a ax + ab = c, where a, b, and c are

    constants.

6. Find the equation of a straight line which is equidistant from the points (2, 3) and (6, 1), expressing it in the form ax + by = c where a, b and c are constants.

7. The equation of a line $^{-3}/_5x + 3y = 6$. Find the:

   (a) Gradient of the line

   (b) Equation of a line passing through point (1, 2) and perpendicular to the given line b

8. Find the equation of the perpendicular to the line x + 2y = 4 and passes through point (2,1)

9. Find the equation of the line which passes through the points P (3,7) and Q (6,1)

10. Find the equation of the line whose x- intercepts is -2 and y- intercepts is 5

11. Find the gradient and y- intercept of the line whose equation is 4x – 3y – 9 = 0

CHAPTER ELEVEN: TRANSFORMATIONS

1. A translation maps a point (1, 2) onto) (-2, 2). What would be the coordinates of the object whose image is (-3, -3) under the same translation?

2. Use binomial expression to evaluate $(0.96)^5$ correct to 4 significant figures

   11. In the figure below triangle ABO represents a part of a school badge. The badge has as symmetry of order 4 about O. Complete the figures to show the badge.

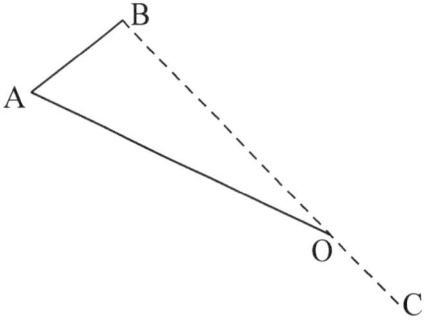

3. A point (-5, 4) is mapped onto (-1, -1) by a translation. Find the image of (-4, 5) under the same translation.

4. A triangle is formed by the coordinates A (2, 1) B (4, 1) and C (1, 6). It is rotated clockwise through 90^0 about the origin. Find the coordinates of this image.

5. The diagram on the grid provided below shows a trapezium ABCD

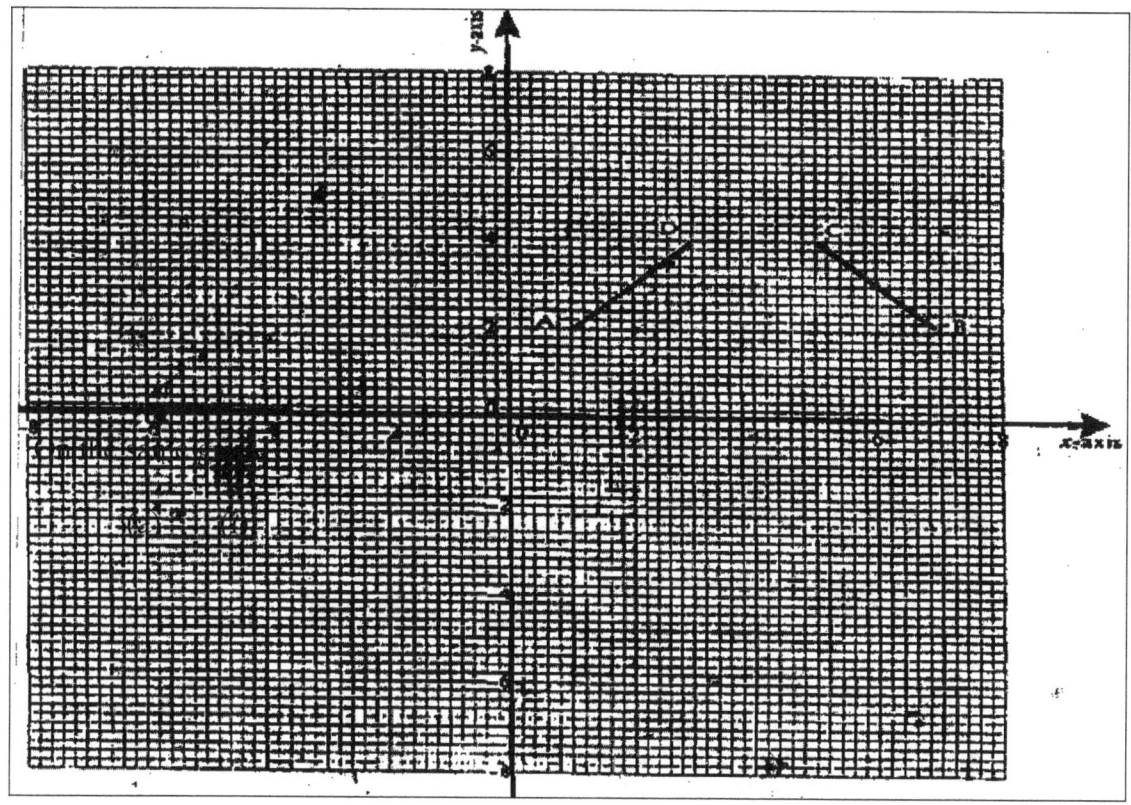

Draw the image A'B'C'D of ABCD under a rotation of 90^0

clockwise about the origin .

   (ii)   Draw the image of A"B"C"D" of A'B'C'D' under a reflection in

          line y = x. State coordinates of A"B"C"D".

   (b)   A"B"C"D" is the image of A"B"C"D under the reflection in the line x=0.

          Draw the image   A"B" C"D" and state its coordinates.

   (c)   Describe a single transformation that maps A" B"C"D onto ABCD.

6. A translation maps a point P(3,2) onto P'(5,4)

   (a) Determine the translation vector

(b) A point Q' is the image of the point Q (, 5) under the same translation. Find the length of 'P' Q leaving the answer is surd form.

7. Two points P and Q have coordinates (-2, 3) and (1, 3) respectively. A translation map point P to P' (10, 10)

   (a) Find the coordinates of Q' the image of Q under the translation

   (b) The position vector of P and Q in (a) above are p and q respectively given that mp – nq = -12 $\begin{pmatrix} \\ \end{pmatrix}$

   9   Find the value of m and n

8. on the Cartesian plane below, triangle PQR has vertices P(2, 3), Q (1,2) and R (4,1) while triangles P" q " R" has vertices P" (-2, 3), Q" (-1,2) and R" (-4, 1)

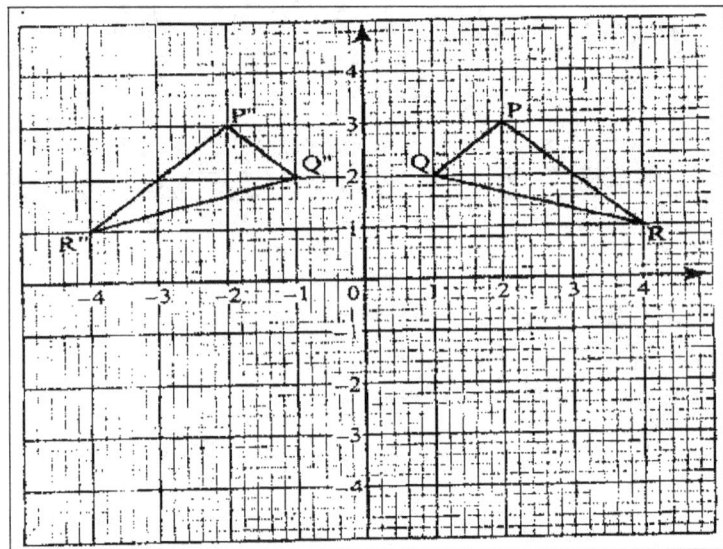

(a)   Describe fully a single transformation which maps triangle PQR onto triangle P"Q"R"

(b)   On the same plane, draw triangle P'Q'R', the image of triangle PQR, under reflection in line y = -x

(c)   Describe fully a single transformation which maps triangle P'Q'R' onto triangle P"Q"R

(d)   Draw triangle P"Q"R" such that it can be mapped onto triangle PQR by a positive quarter turn about (0, 0)

46

(e)   State all pairs of triangle that are oppositely congruent

CHAPTER TWELVE: MEASUREMENTS II

1. A solid cone of height 12 cm and radius 9 cm is recast into a solid sphere. Calculate the surface area of the sphere.

2. A circular path of width 14 metres surrounds a field of diameter 70 metres. The path is to be carpeted and the field is to have a concrete slab with an exception of four rectangular holes each measuring 4 metres by 3 metres.

   A contractor estimated the cost of carpeting the path at Kshs. 300 per square metre and the cost of putting the concrete slab at Kshs 400 per square metre. He then made a quotation which was 15% more than the total estimate. After completing the job, he realized that 20% of the quotation was not spent.

   (a) How much money was not spent?

   (b) What was the actual cost of the contract?

3. In the figure below BAD and CBD are right angled triangles.

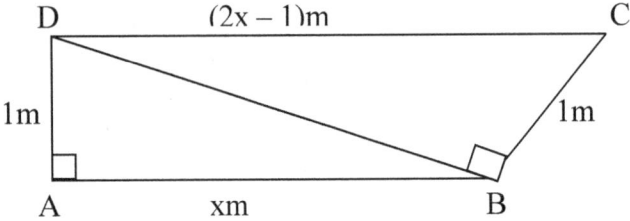

   Find the length of AB.

4. A cylinder of radius 14 cm contains water. A metal solid cone of base radius 7 cm and height 18 cm is submerged into the water. Find the change in height of the water level in the cylinder.

5. A cyndrical container of radius 15 cm has some water in it. When a solid is submerged into the water, the water level rises by 1.2 cm.

   (a) Find, the volume of the water displaced by the solid leaving your answer in terms of

   ∏

   (b) If the solid is a circular cone of height 9 cm, calculate the radius of the cone to 2 decimal places.

6. A balloon, in the form of a sphere of radius 2 cm, is blown up so that the volume increases by 237.5%. Determine the new volume of balloon in terms of ∏

A girl wanted to make a rectangular octagon of side 14 cm. She made it from a square piece of a card of size y cm by cutting off four isosceles triangles whose equal sides were x cm each, as shown below.

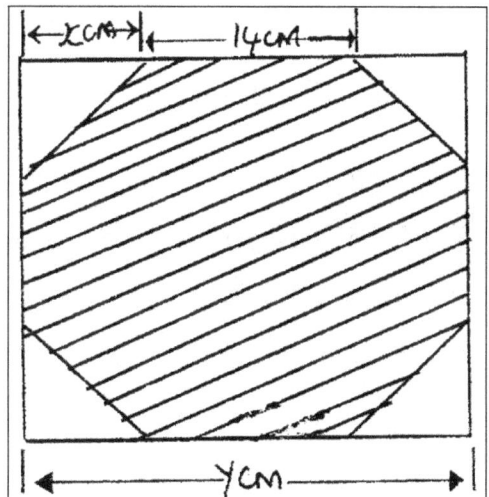

   (a) Write down an expression for the octagon in terms of x and y

   (b) Find the value of x

   (c) Find the area of the octagon

7. A pyramid VABCD has a rectangular horizontal base ABCD with AB= 12 cm and BC = 9 cm. The vertex V is vertically above A and VA = 6 cm. calculate the volume of the pyramid.

8. A solid made up of a conical frustrum and a hemisphere top as shown in the figure below.

   The dimensions are as indicated in the figure.

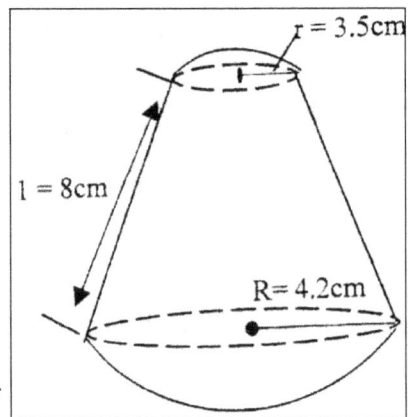

   (a) Find the area of

      (i)   The circular base

      (ii)   The curved surface of the frustrum

      (iii)   The hemisphere surface

   (b) A similar solid has a total area of 81.51 cm². Determine the radius of its base.

9. Two sides of triangles are 5 cm each and the angle between them is 120^0. Calculate the area

   of the triangle.

10. The figure below represents a kite ABCD, AB = AD = 15 cm. The diagonals BD and AC

    intersect at O. AC = 30 cm and AO = 12 cm.

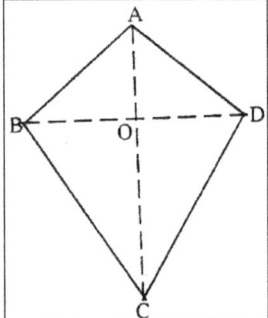

Find the area of the kite

11. The diagram below represents a solid made up of a hemisphere mounted on a

cone. The radius of the hemisphere are each 6 cm and the height of the cone is 9 cm.

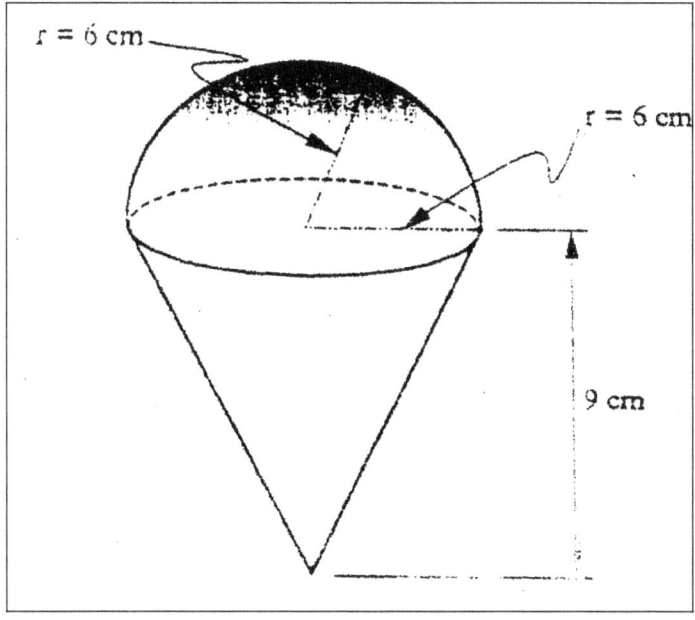

Calculate the volume of the solid take п = ($^{22}/_7$)

12. The internal and external diameters of a circular ring are 6 cm and 8cm respectively. Find the

volume of the ring if its thickness is 2 millimeters.

2003

13. A wire of length 21 cm is bent to form the shape down in the figure below, ABCD is a

rectangle and AEB is an equilateral triangle.

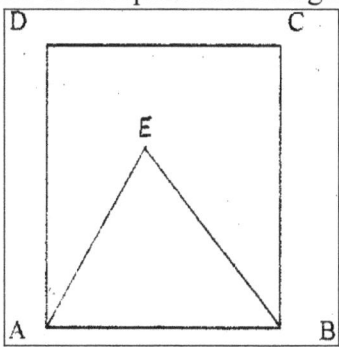

If the length of AD of the rectangle is 1 ½ times its width, calculate the

width of the rectangle.

14. The length of a hallow cylindrical pipe is 6 metres. Its external diameter is 11 cm and has a thickness of 1 cm. Calculate the volume in cm³ of the material used to make the pipe. Take π as 3.142.

15. The figure below represents a hexagon of side 5 cm and 20 cm height

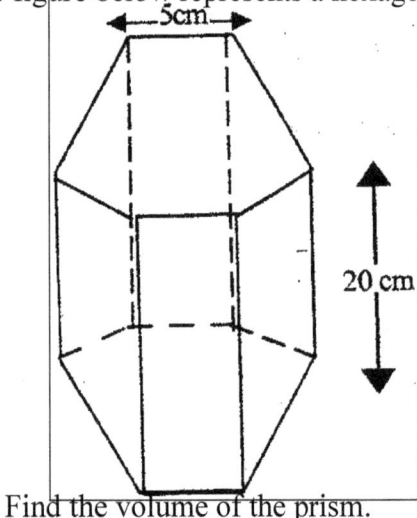

Find the volume of the prism.

16. A cylindrical piece of wood of radius 4.2 cm and length 150 cm is cut length into two equal pieces.

Calculate the surface area of one piece

(Take π as $^{22}/_7$

17. The figure below is a model representing a storage container. The model whose total height is 15 cm is made up of a conical top, a hemispherical bottom and the middle part is cylindrical. The radius of the base of the cone and that of the hemisphere are each 3 cm. The height of the cylindrical part is 8cm.

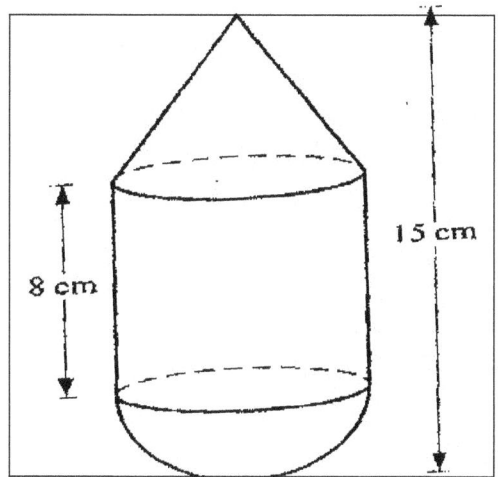

(a) Calculate the external surface area of the model

    (b)   The actual storage container has a total height of 6 metres. The outside of the actual storage container is to be painted. Calculate the amount of paint required if an area of $20m^2$ requires 0.75 litres of the paint.

18. A garden measures 10m long and 8 m wide. A path of uniform width is made all round the garden. The total area of the garden and the paths is 168 m^2.

(a) Find the width of the path

(b) The path is to be covered with square concrete slabs. Each corner of the path is covered with a slab whose side is equal to the width of the path.

    The rest of the path is covered with slabs of side 50 cm. The cost of making each corner slab is Kshs 600 while the cost of making each smaller slab is Kshs 50.

    Calculate

        (i) The number of smaller slabs used

        (ii) The total cost of the slabs used to cover the whole path

19. A cylindrical solid of radius 5 cm and length 12 cm floats lengthwise in water to a depth of 2.5 cm as shown in the figure below.

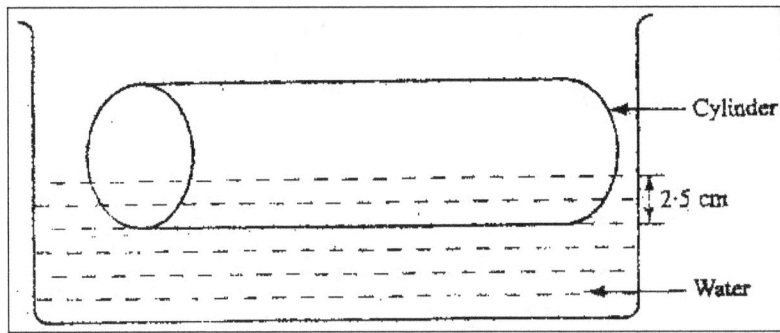

Calculate the area of the curved surface of the solid in contact with water, correct to 4 significant figures

20. Two cylindrical containers are similar. The larger one has internal cross- section area of 45 cm^2 and can hold 0.945 litres of liquid when full. The smaller container has internal cross-section area of 20 cm^2

(a) Calculate the capacity of the smaller container

(b) The larger container is filled with juice to a height of 13 cm. Juice is then drawn from is and emptied into the smaller container until the depths of the juice in both containers are equal.

Calculate the depths of juice in each container.

(c) On fifth of the juice in the larger container in part (b) above is further drawn and emptied into the smaller container. Find the difference in the depths of the juice in the two containers.

22. A metal bar is a hexagonal prism whose length is 30 cm, the cross section is a regular hexagon with each side of length 6 cm

    Find

    (i)   The area of the hexagonal face

    (ii)   The volume of the metal bar

23. A cylindrical water tank of diameter 7 metres and height 2.8 metres

    (a)   Find the capacity of the water tank in litres

    (b)   Six members of family use 15 litres each per day. Each day 80 litres are sued for cooking and washing and a further 60 litrese are wasted.

            Find the number of complete days a full tank would last the family (2mks)

    (c)   Two members of the family were absent for 90 days. During the 90 days wastage was reduced by 20% but cooking and washing remained the same. Calculate the number of days a full tank would now last the family

24. Pieces of soap are packed in a cuboid container measuring 36 cm by 24 cm by 18 cm. Each piece of soap is a similar to the container. If the linear scale factor between the container and the soap is $\frac{1}{6}$ find the volume of each piece of soap.

25. A pyramid of height 10cm stands, on a square base ABCD of side 6 cm.

    (a)   Draw a sketch of the pyramid

    (b)   Calculate the perpendicular distance from the vertex to the side AB.

CHAPTER THIRTEEN: QUADRATIC EQUATIONS

1. Simplify

$$\frac{2x - 2}{6x^2 - x - 12} \div \frac{x - 1}{2x - 3}$$

2. Solve the simultaneous equations

$$2x - y = 3$$

$$x^2 - xy = -4$$

3. Find the value of x in the following equations:

$$49^{x + 1} + 7^{2x} = 350$$

4. Simplify completely

$$\frac{3x^2 - 1}{x^2 - 1} - \frac{2x + 1}{x + 1}$$

5. Factorize completely $3x^2 - 2xy - y^2$

6. Factorize $a^2 - b^2$

   Hence find the exact value of $2557^2 - 2547^2$

7. If $x^2 + y^2 = 29$ and $x + y = 3$

   (a) Determine the values of

       (i) $x^2 + 2xy + y^2$

       (ii) $2xy$

       (iii) $x^2 - 2xy + y^2$

       (iv) $x - y$

   (b) Find the value of x and y

8. Simplify the expression $\dfrac{3a^2 + 4ab + b^2}{4a^2 + 3ab - b^2}$

9. (a) Write an expression in terms of x and y for the total value of a two digit number having x as the tens digit and y as the units digit.

   b) The number in (a) above is such that three times the sum of its digits is less than the value of the number by 8. When the digits are reversed the value of the number increases by 9. Find the number xy.

10. Simplify the expression $\dfrac{2a^2 - 3\,ab - 2b^2}{4a^2 - b^2}$

11. Simplify the expression $\dfrac{9t^2 - 25a^2}{6t^2 + 19\,at + 15a^2}$

12. Simplify

$$\dfrac{P^2 + 2pq + q^2}{P^3 - pq^2 + p^2q - q^3}$$

13. Expand the expression $(x^2 - y^2)(x^2 + y^2)(x^4 - y^4)$

14. The sum of two numbers x and y is 40. Write down an expression, in terms of x, for the sum of the squares of the two numbers.

    Hence determine the minimum value of $x^2 + y^2$

15. Simplify the expression $\dfrac{15a^2b - 10ab^2}{-5ab + 2b^2}$

16. Four farmers took their goats to the market Mohamed had two more goats than Ali Koech had 3 times as many goats as Mohamed. Whereas Odupoy had 10 goats less than both Mohamed and Koech.

(i) Write a simplified algebraic expression with one variable. Representing the total number of goats

(ii) Three butchers bought all the goats and shared them equally. If each butcher got 17 goats. How many did Odupoy sell to the butchers?

17. Find the value of x which satisfies the equation $16^{2x} = 8^{4x-3}$

18. Mary has 21 coins whose total value is Kshs 72. There are twice as many five shillings coins as there are ten shillings coins. The rest one shilling coins. Find the number of ten shilling coins that Mary has.

19. Simplify the expression

$$\frac{x-1}{x} - \frac{2x+1}{3x}$$

Hence solve the equation $\frac{x-1}{x} - \frac{2x+1}{3x} = \frac{2}{3}$

20. Given that $P = 3^y$ express the equation

$3^{2y-1} + 2 \times 3^{y-1} = 1$ in terms of P.

Hence or otherwise find the value of y in the equation $3^{2y-1} + 2 \times 3^{y-1} = 1$

21. Simplify the expression

$$\frac{4x^2 - y^2}{2x^2 - 7xy + 3y}$$

22. Three years ago Juma was three times as old as Ali. In two years time the sum of their ages will be 62. Determine their present ages.

23. Simplify

$$\frac{x - 2}{x + 2} - \frac{2x + 20}{x^2 - 4}$$

24. If the expression $25y^2 - 70y + d$ is a perfect square, where d is a constant, find the value of d

CHAPTER FOURTEEN: INEQUALITIES

1. Find the range of x if $2 \leq 3 - x < 5$

2. Find all the integral values of x which satisfy the inequalities:

   $2(2-x) < 4x - 9 < x + 11$

3. Solve the inequality and show the solution

   $3 - 2x \angle x \leq \dfrac{2x + 5}{3}$ on the number line

4. Solve the inequality $\dfrac{x - 3}{4} + \dfrac{x - 5}{6} \leq \dfrac{4x + 6}{8} - 1$

5. A family is planning a touring holiday, during which time (x days) will be spent walking and the rest of the time (y days) in traveling by bus. Each day they can walk 30 km or travel 80 km by bus and they wish to travel at least 600 km altogether.

   The holiday must not last more than 14 days. Each day walking will cost Kshs. 200 and each day traveling by bus will cost Kshs. 1400. The holiday must not cost more than Kshs 9800

   (a) Write down all the inequalities in x and y based on the above facts

   (b) Represent the inequalities graphically

   (c) Use the graph to determine the integral values of x and y which give

      (i) The cheapest holiday

      (ii) The longest distance traveled

CHAPTER FIFTEEN: CIRCLES

1. In the figure below CP= CQ and <CQP = 160⁰. If ABCD is a cyclic quadrilateral, find < BAD.

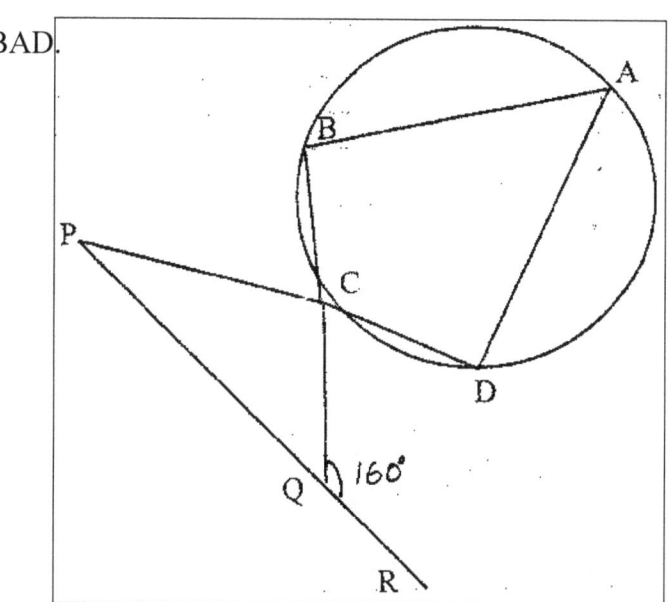

2. In the figure below AOC is a diameter of the circle centre O; AB = BC and < ACD = 25⁰, EBF is a tangent to the circle at B.G is a point on the minor arc CD.

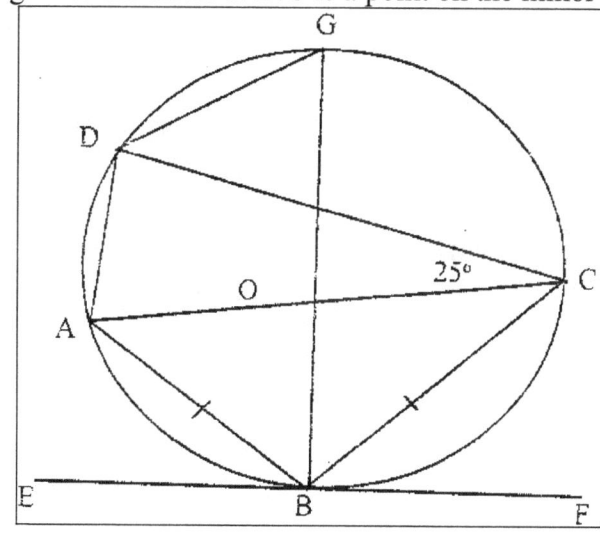

(a) Calculate the size of

    (i) < BAD

    (ii) The Obtuse < BOD

(iii) < BGD

(b)   Show the < ABE = < CBF. Give reasons

3. In the figure below PQR is the tangent to circle at Q. TS is a diameter and TSR and QUV are straight lines. QS is parallel to TV. Angles SQR = 40^0 and angle TQV = 55^0

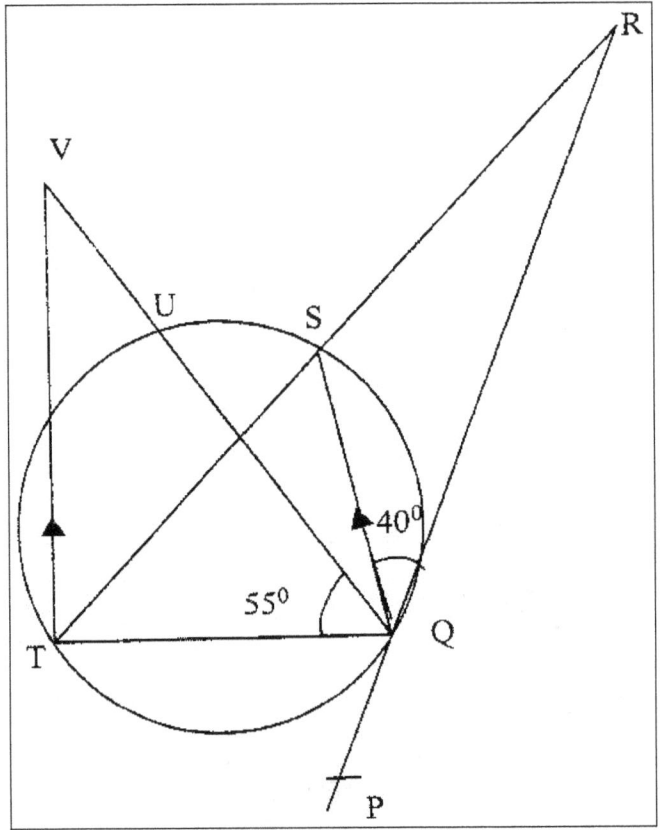

Find the following angles, giving reasons for each answer

(a)   QST

(b)   QRS

(c)   QVT

(d)   UTV

4.

$= 37^0$

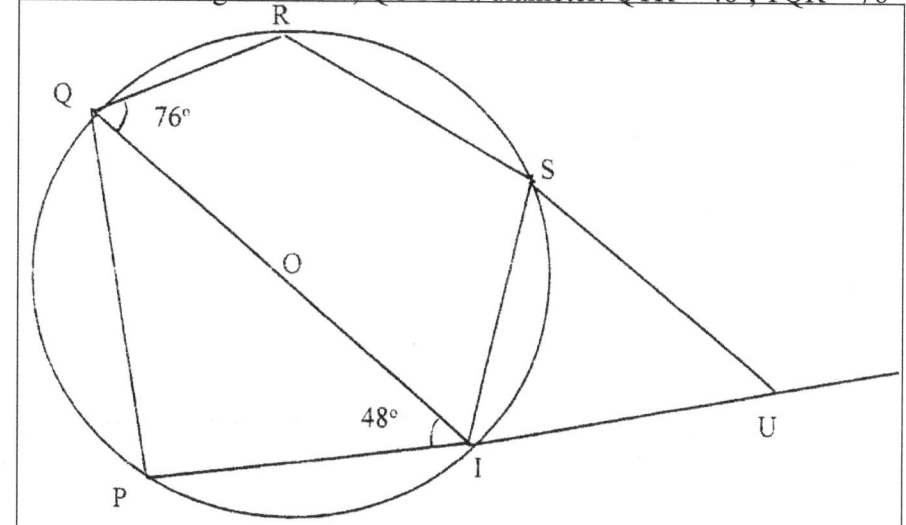

In the figure below, QOT is a diameter. QTR = 48^0, TQR = 76^0 and SRT

Calculate

   (a)   <RST

   (b)   <SUT

   (c)   Obtuse <ROT

5.   In the figure below, points O and P are centers of intersecting circles ABD and

       BCD respectively. Line ABE is a tangent to circle BCD at B. Angle BCD = 42^0

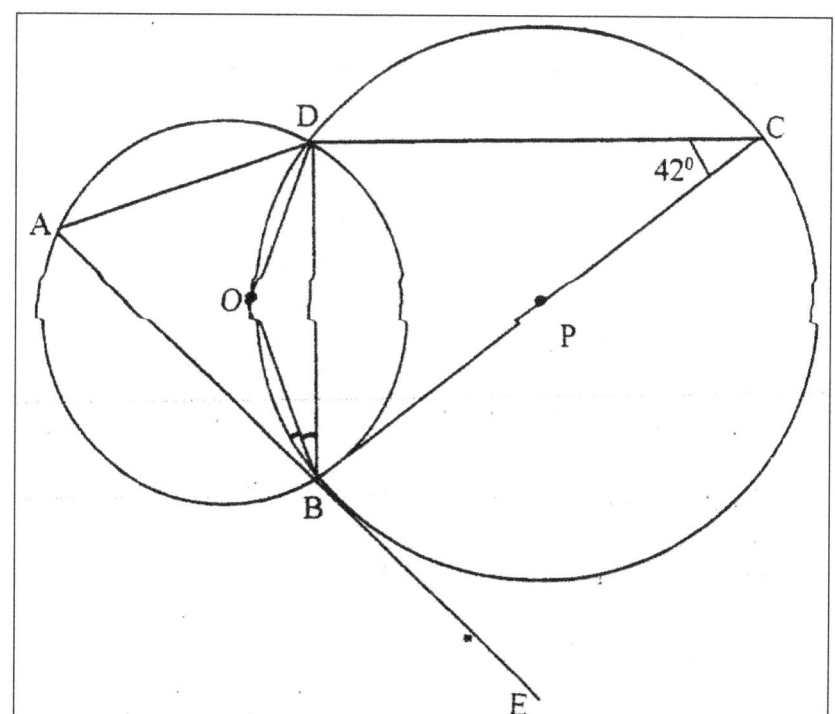

(a)   Stating reasons, determine the size of

    (i)   <CBD

    (ii) Reflex <BOD

 (b) Show that Δ ABD is isosceles

6. The diagram below shows a circle ABCDE. The line FEG is a tangent to the circle at

  point E. Line DE is parallel to CG, <DEC = 28^0 and < AGE = 32^0

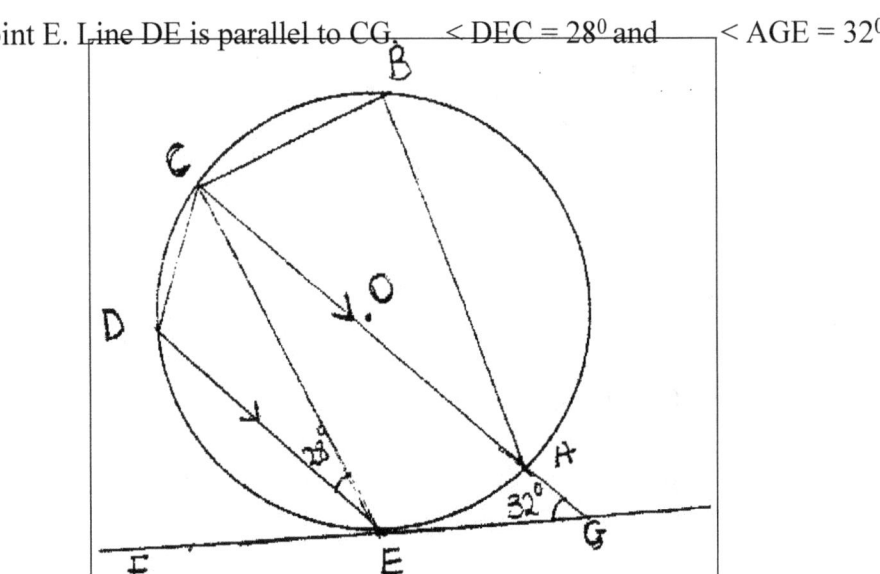

Calculate:

  (a) < AEG

  (b) < ABC

7.   In the figure below R, T and S are points on a circle centre O PQ is a tangent to

the circle at T. POR is a straight line and ∠ QPR = 20⁰

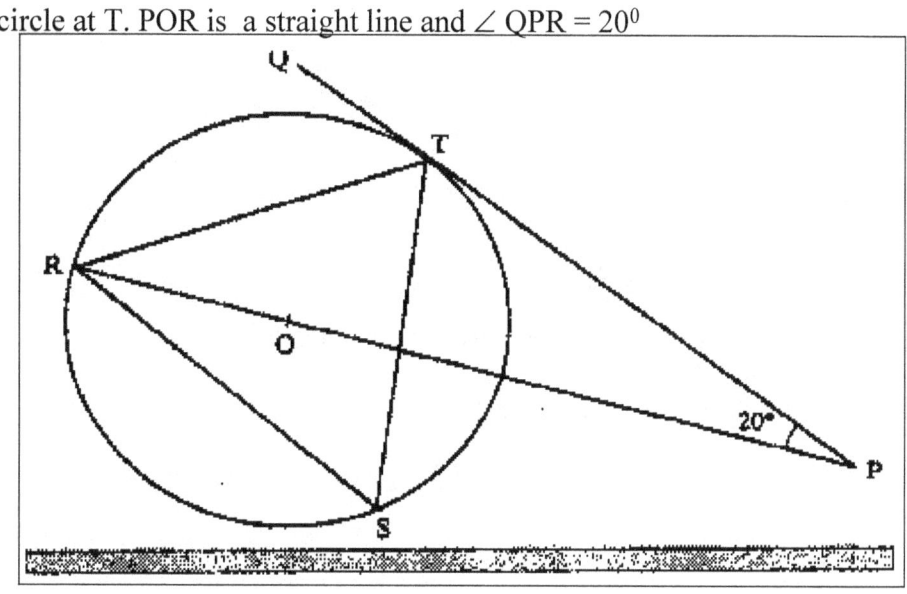

Find the size of ∠ RST

8.   The figure below shows a circle centre O and a point Q which is outside

the

circle

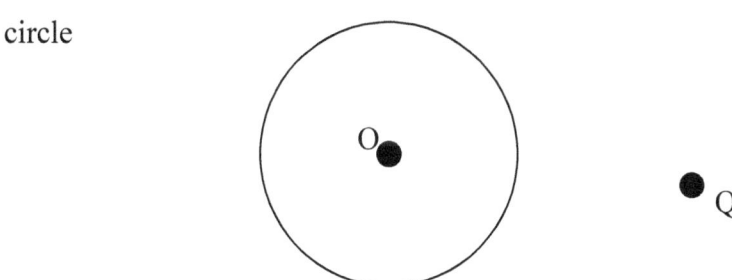

Using a ruler and a pair of compasses, only locate a point on the circle such that angle OPQ = 90º

CHAPTER SIXTEEN: LINEAR MOTION

1. Two towns P and Q are 400 km apart. A bus left P for Q. It stopped at Q for one hour and then started the return journey to P. One hour after the departure of the bus from P, a trailer also heading for Q left P. The trailer met the returning bus ¾ of the way from P to Q. They met t hours after the departure of the bus from P.

   (a) Express the average speed of the trailer in terms of t

   (b) Find the ration of the speed of the bus so that of the trailer.

2. The athletes in an 800 metres race take 104 seconds and 108 seconds respectively to complete the race. Assuming each athlete is running at a constant speed. Calculate the distance between them when the faster athlete is at the finishing line.

3. A and B are towns 360 km apart. An express bus departs form A at 8 am and maintains an average speed of 90 km/h between A and B. Another bus starts from B also at 8 am and moves towards A making four stops at four equally spaced points between B and A. Each stop is of duration 5 minutes and the average speed between any two spots is 60 km/h. Calculate distance between the two buses at 10 am.

4. Two towns A and B are 220 km apart. A bus left town A at 11. 00 am and traveled towards B at 60 km/h. At the same time, a matatu left town B for town A and traveled at 80 km/h. The matatu stopped for a total of 45 minutes on the way before meeting the bus. Calculate the distance covered by the bus before meeting the matatu.

5. A bus travels from Nairobi to Kakamega and back. The average speed from Nairobi to Kakamega is 80 km/hr while that from Kakamega to Nairobi is 50 km/hr, the fuel consumption is 0.35 litres per kilometer and at 80 km/h, the consumption is 0.3 litres per kilometer .Find

i)   Total fuel consumption for the round trip

ii)   Average fuel consumption per hour for the round trip.

6.   The distance between towns M and N is 280 km. A car and a lorry travel from M to N. The average speed of the lorry is 20 km/h less than that of the car. The lorry takes 1h 10 min more than the car to travel from M and N.

(a)   If the speed of the lorry is x km/h, find x

(b)   The lorry left town M at 8: 15 a.m. The car left town M and overtook the lorry at 12.15 p.m. Calculate the time the car left town M.

7.   A bus left Mombasa and traveled towards Nairobi at an average speed of 60 km/hr. after 21/2 hours; a car left Mombasa and traveled along the same road at an average speed of 100 km/ hr. If the distance between Mombasa and Nairobi is 500 km, Determine

(a)   (i)   The distance of the bus from Nairobi when the car took off

(ii)   The distance the car traveled to catch up with the bus

(b)   Immediately the car caught up with the bus

(c)   The car stopped for 25 minutes. Find the new average speed at which the car traveled in order to reach Nairobi at the same time as the bus.

8.   A rally car traveled for 2 hours 40 minutes at an average speed of 120 km/h. The car consumes an average of 1 litre of fuel for every 4 kilometers.

A litre of the fuel costs Kshs 59

Calculate the amount of money spent on fuel

9. A passenger notices that she had forgotten her bag in a bus 12 minutes after the bus had left. To catch up with the bus she immediately took a taxi which traveled at 95 km/hr. The bus maintained an average speed of 75 km/ hr. determine

(a) The distance covered by the bus in 12 minutes

(b) The distance covered by the taxi to catch up with the bus

10. The athletes in an 800 metre race take 104 seconds and 108 seconds respectively to complete the race. Assuming each athlete is running at a constant speed. Calculate the distance between them when the faster athlete is at the finishing line.

11. Mwangi and Otieno live 40 km apart. Mwangi starts from his home at 7.30 am and cycles towards Otieno's house at 16 km/ h Otieno starts from his home at 8.00 and cycles at 8 km/h towards Mwangi at what time do they meet?

12. A train moving at an average speed of 72 km/h takes 15 seconds to completely cross a bridge that is 80m long.

(a) Express 72 km/h in metres per second

(b) Find the length of the train in metres

CHAPTER SEVENTEEN: QUADRATIC EXPRESSIONS AND EQUATIONS

1. The table shows the height metres of an object thrown vertically upwards varies with the time t seconds

    The relationship between s and t is represented by the equations $s = at^2 + bt + 10$ where b are constants.

T	0	1	2	3	4	5	6	7	8	9	10
S		45.1									

    (a) (i) Using the information in the table, determine the values of a and b

        (ii) Complete the table

    (b)(i) Draw a graph to represent the relationship between s and t

        (ii) Using the graph determine the velocity of the object when t = 5 seconds

2. (a) Construct a table of value for the function $y = x^2 - x - 6$ for $-3 \leq x \leq 4$

    (b) On the graph paper draw the graph of the function

        $Y = x^2 - x - 6$ for $-3 \leq x \leq 4$

    (c) By drawing a suitable line on the same grid estimate the roots of the equation

        $x^2 + 2x - 2 = 0$

3. (a) Draw the graph of $y = 6 + x - x^2$, taking integral value of x in $-4 \leq x \leq 5$. (The grid is provided. Using the same axes draw the graph of $y = 2 - 2x$

    (b) From your graphs, find the values of X which satisfy the simultaneous equations $y = 6 + x - x^2$

$$y = 2 - 2x$$

(c) Write down and simplify a quadratic equation which is satisfied by the values of x where the two graphs intersect.

4. (a) Complete the following table for the equation $y = x^3 - 5x^2 + 2x + 9$

X	-2	-1.5	-1	0	1	2	3	4	5
x^2		-3.4	-1	0	1		27	64	125
$-5x^2$	-20	-11.3	-5	0	-1	-20	-45		
$2x$	-4	-3		0	2	4	6	8	10
9	9	9	9	9	9	9	9	9	99
		-8.7			9	7		-3	

(b) On the grid provided draw the graph of $y = x^3 - 5x^2 + 2x + 9$ for $-2 \leq x \leq 5$

(c) Using the graph estimate the root of the equation $x^3 - 5x^2 + 2 + 9 = 0$ between x = 2 and x = 3

(d) Using the same axes draw the graph of $y = 4 - 4x$ and estimate a solution to the equation $x^2 - 5x^2 + 6x + 5 = 0$

5. (a) Complete the table below, for function $y = 2x^2 + 4x - 3$

X	-4	-3	-2	-1	0	1	2
$2x^2$	32		8	2	0	2	
$4x - 3$			-11		-3		5
Y			-3			3	13

(b) On the grid provided, draw the graph of the function $y = 2x^2 + 4x - 3$ for

$-4 \leq x \leq 2$ and use the graph to estimate the rots of the equation $2x^2+4x-3=0$ to 1 decimal place.

(c)   In order to solve graphically the equation $2x^2 +x -5 =0$, a straight line must be drawn to intersect the curve $y = 2x^2 + 4x - 3$. Determine the equation of this straight line, draw the straight line hence obtain the roots.

$2x^2 + x - 5$ to 1 decimal place.

6.   (a)   (i)   Complete the table below for the function $y = x^3 + x^2 - 2x$ (2mks)

X	-3	-2.5	-2	-1.5	-1	-0.5	0	0.5	1	2	2.5
x^3		15.63				-0.13			1		
x^2			4					0.25			6.25
-2x						1			-2		
y				1.87				0.63			16.88

(ii)   On the grid provided, draw the graph of $y = x^3 + x^2 - 2x$ for the values of x in the interval $-3 \leq x \leq 2.5$

(iii)   State the range of negative values of x for which y is also negative

(b)   Find the coordinates of two points on the curve other than (0, 0) at which x-coordinate and y- coordinate are equal

7.   The table shows some corresponding values of x and y for the curve represented by $Y = \frac{1}{4}x3 -2$

X	-3	-2	-1	0	1	2	3
Y	-8.8	-4	-2.3	-2	-1.8	0	4.8

On the grid provided below, draw the graph of $y = \frac{1}{4} x^2 - 2$ for $-3 \leq x \leq 3$. Use the graph to estimate the value of x when y = 2

8. A retailer planned to buy some computers form a wholesaler for a total of Kshs 1,800,000. Before the retailer could buy the computers the price per unit was reduced by Kshs 4,000. This reduction in price enabled the retailer to buy five more computers using the same amount of money as originally planned.

(a) Determine the number of computers the retailer bought

(b) Two of the computers purchased got damaged while in store, the rest were sold and the retailer made a 15% profit Calculate the profit made by the retailer on each computer sold

9. The figure below is a sketch of the graph of the quadratic function y = k

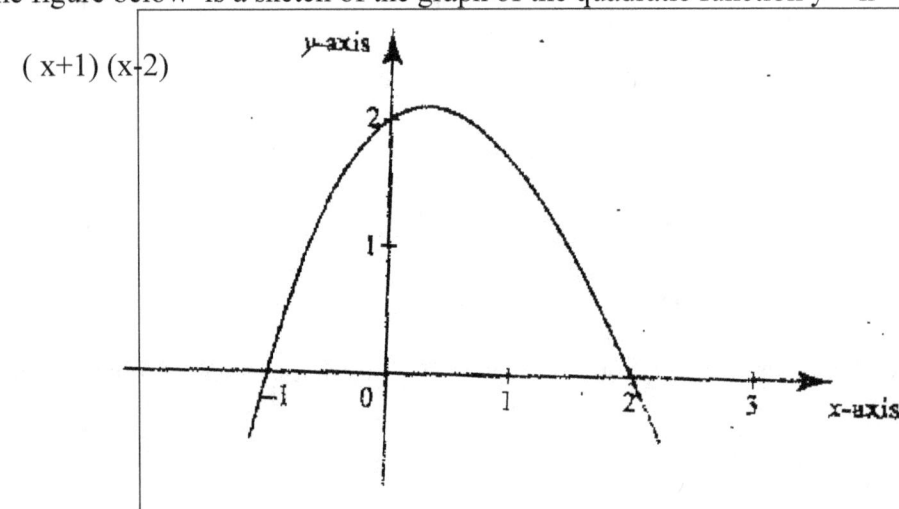

(x+1) (x-2)

Find the value of k

10. (a) Draw the graph of $y = x^2 - 2x + 1$ for values $-2 \leq x \leq 4$

(b) Use the graph to solve the equations $x^2 - 4 = 0$ abd line $y = 2x + 5$

11.   (a)   Draw the graph of $y = x^3 + x^2 - 2x$ for $-3 \leq x \leq 3$ take scale of 2cm to

represent 5 units as the horizontal axis

   (b)   Use the graph to solve $x^3 + x^2 - 6 - 4 = 0$ by drawing a suitable linear graph on the

same axes.

12.   Solve graphically the simultaneous equations $3x - 2y = 5$ and $5x + y = 17$

CHAPTER EIGHTEEN: APPROXIMATION AND ERRORS

1.   (a) Work out the exact value of $R = \dfrac{1}{0.003146 - 0.003130}$

   (b)   An approximate value of R may be obtained by first correcting each of the

       decimal in the denominator to 5 decimal places

       (i)   The approximate value

       (ii)   The error introduced by the approximation

2.   The radius of circle is given as 2.8 cm to 2 significant figures

   (a) If C is the circumference of the circle, determine the limits between which C/π lies

   (b) By taking $\prod$ to be 3.142, find, to 4 significant figures the line between which the

       circumference lies.

3.   The length and breath of a rectangular floor were measured and found to be 4.1 m and 2.2

   m respectively. If possible error of 0.01 m was made in each of the measurements, find

   the:

       (a) Maximum and minimum possible area of the floor

       (b) Maximum possible wastage in carpet ordered to cover the whole floor

4.   In this question Mathematical Tables should not be used

   The base and perpendicular height of a triangle measured to the nearest centimeter

   are 6 cm and 4 cm respectively.

   Find

   (a) The absolute error in calculating the area of the triangle

   (b) The percentage error in the area, giving the answer to 1 decimal place

5. By correcting each number to one significant figure, approximate the value of 788 x 0.006. Hence calculate the percentage error arising from this approximation.

6. A rectangular block has a square base whose side is exactly 8 cm. Its height measured to the nearest millimeter is 3.1 cm

   Find in cubic centimeters, the greatest possible error in calculating its volume.

7. Find the limits within the area of a parallegram whose base is 8cm and height is 5 cm lies. Hence find the relative error in the area

8. Find the minimum possible perimeter of a regular pentagon whose side is 15.0cm.

9. Given the number 0.237

   (i)   Round off to two significant figures and find the round off error

   (ii)   Truncate to two significant figures and find the truncation error

10. The measurements a = 6.3, b= 15.8, c= 14.2 and d= 0.00173 have maximum possible errors of 1%, 2%, 3% and 4% respectively. Find the maximum possible percentage error in $^{ad}/_{bc}$ correct to 1sf.

CHAPTER NINETEEN: TRIGONOMETRY 1

1. Solve the equation

    $$\text{Sin } \frac{5}{2}\,\theta = \frac{-1}{2} \text{ for } 0^0 \le 0 \le 180^0$$

2. Given that $\sin \theta = {}^2/_3$ and is an acute angle find:

    (a) Tan θ giving your answer in surd form

    (b) Sec^2 θ

3. Solve the equation $2 \sin^2(x-30^0) = \cos 60^0$ for $-180^0 \le x \le 180^0$

4. Given that $\sin (x + 30)^0 = \cos 2x^0$ for 0^0, $0^0 \le x \le 90^0$ find the value of x. Hence find the

    value of $\cos {}^23x^0$.

5. Given that $\sin a = \frac{1}{\sqrt{5}}$ where a is an acute angle find, without using

    Mathematical tables

    (a) Cos a in the form of a$\sqrt{b}$, where a and b are rational numbers

    (b) Tan $(90^0 - a)$.

6. Give that x^o is an angle in the first quadrant such that $8 \sin^2 x + 2 \cos x -5=0$

    Find:

    a) Cos x

    b) tan x

7. Given that $\cos 2x^0 = 0.8070$, find x when $0^0 \le x \le 360^0$

76

8   The figure below shows a quadrilateral ABCD in which AB = 8 cm, DC = 12 cm, < BAD

= 45^0, < CBD = 90^0 and BCD = 30^0.

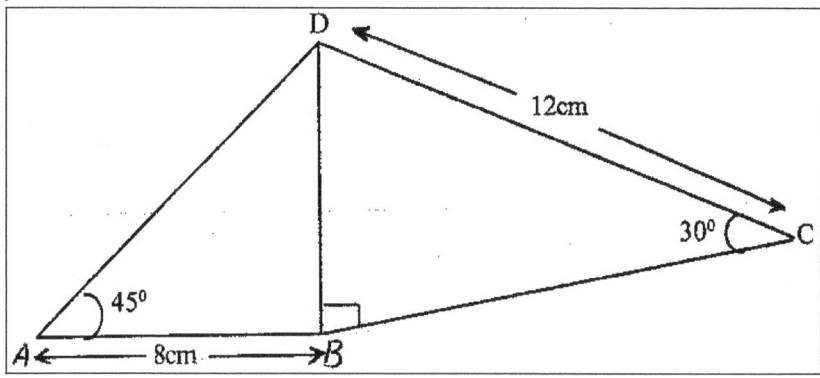

Find:

(a)   The length of BD

(b)   The size of the angle ADB

9.   The diagram below represents a school gate with double shutters. The shutters are such

opened through an angle of 63^0.

The edges of the gate, PQ and RS are each 1.8 m

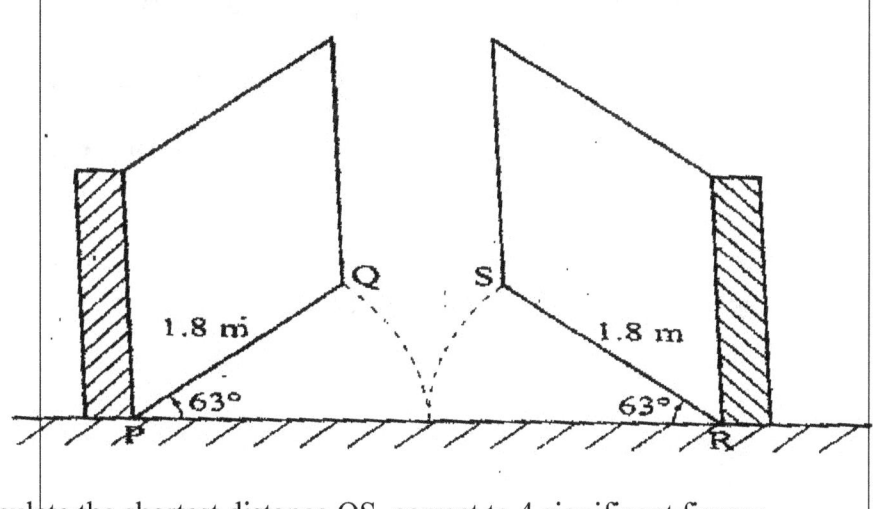

Calculate the shortest distance QS, correct to 4 significant figures

10. The figure below represents a quadrilateral piece of land ABCD divided into three

triangular plots. The lengths BE and CD are 100m and 80m respectively. Angle ABE = 30⁰

$\angle$ACE = 45⁰ and $\angle$ ACD = 100⁰

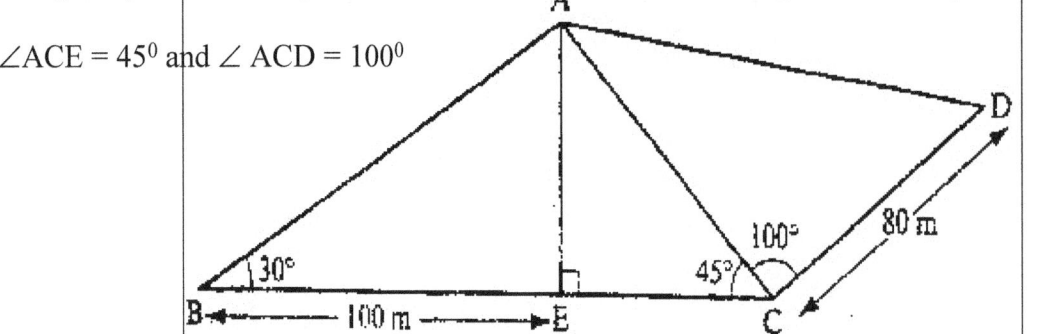

(a) Find to four significant figures:

   (i)   The length of AE

 (ii)   The length of AD

   (iii)   The perimeter of the piece of land

(b) The plots are to be fenced with five strands of barbed wire leaving an entrance of 2.8 m

wide to each plot. The type of barbed wire to be used is sold in rolls of lengths 480m.

Calculate the number of rolls of barbed wire that must be bought to complete the fencing

of the plots.

11. Given that x is an acute angle and cos x = $\frac{2\sqrt{5}}{5}$, find without using mathematical

tables or a calculator, tan (90 – x)⁰.

12. In the figure below $\angle A = 62^0$, $\angle B = 41^0$, BC = 8.4 cm and CN is the bisector of $\angle ACB$.

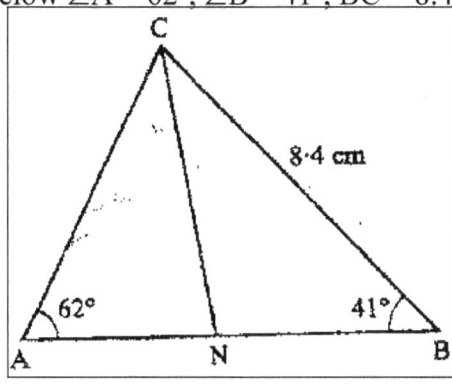

Calculate the length of CN to 1 decimal place.

13. In the diagram below PA represents an electricity post of height 9.6 m. BB and RC represents two storey buildings of heights 15.4 m and 33.4 m respectively. The angle of depression of A from B is 5.5^0 While the angle of elevation of C from B is 30.5^0 and BC = 35m.

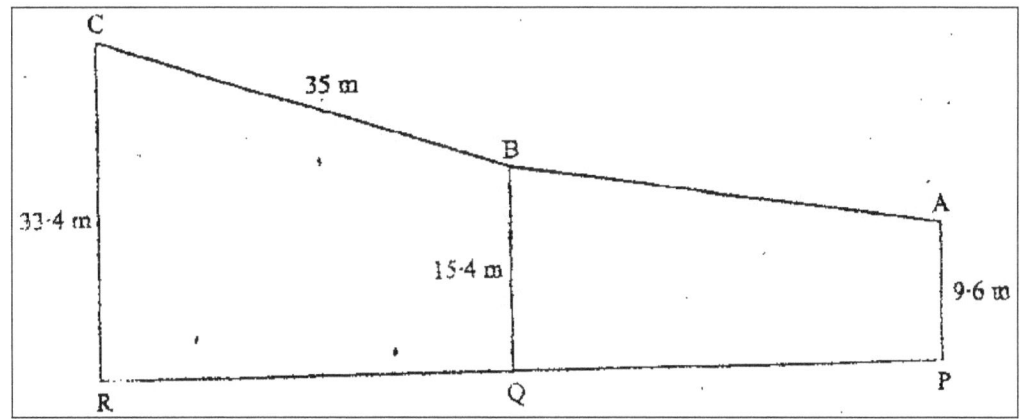

   (a)   Calculate, to the nearest metre, the distance AB

   (b)   By scale drawing find,

      (i)   The distance AC in metres

      (ii) $\angle$ BCA and hence determine the angle of depression of A from C

CHAPTER TWENTY: SURDS AND FURTHER LOGARITHM

1. Without using logarithm tables, find the value of x in the equation

    $\log x^3 + \log 5x = 5 \log 2 - \log \underline{2}$

    $\qquad\qquad\qquad\qquad\qquad\qquad\qquad 5$

2. Simplify $\qquad (1 \div \sqrt{3})(1 - \sqrt{3})$

    Hence evaluate $\qquad \dfrac{1}{1 + \sqrt{3}}$ to 3 s.f. given that $\sqrt{3} = 1.7321$

3. If $\dfrac{\sqrt{14}}{\sqrt{7} - \sqrt{2}} - \dfrac{\sqrt{14}}{\sqrt{7} + \sqrt{2}} = a\sqrt{7} + b\sqrt{2}$

    Find the values of a and b where a and b are rational numbers.

4. Find the value of x in the following equation $49^{(x+1)} + 7^{(2x)} = 350$

5. Find x if $3 \log 5 + \log x^2 = \log 1/125$

6. Simplify as far as possible leaving your answer inform of a surd

    $\dfrac{1}{\sqrt{14} - 2\sqrt{3}} - \dfrac{1}{\sqrt{14} + 2\sqrt{3}}$

7. Given that $\tan 75^0 = 2 + \sqrt{3}$, find without using tables $\tan 15^0$ in the form $p + q\sqrt{m}$, where p, q and m are integers.

8. Without using mathematical tables, simplify

    $\dfrac{\sqrt{63} + \sqrt{72}}{\sqrt{32} + \sqrt{28}}$

9. Simplify $\qquad \dfrac{3}{\sqrt{5} - 2} + \dfrac{1}{\sqrt{5}}$ leaving the answer in the form $a + b\sqrt{c}$, where a, b and c

are rational numbers

10. Given that $P = 3^y$ express the questions $3^{2y-1)} + 2 \times 3^{(y-1)} = 1$ in terms of P

Hence or otherwise find the value of y in the equation: $3^{(2y-1)} + 2 \times 3^{(y-1)} = 1$

11. Solve for $(\log^3 x)^2 - \tfrac{1}{2} \log_3 x = 3/2$

12. Find the values of x which satisfy the equation $5^{2x} - 6(5^x) + 5 = 0$

13. Solve the equation

$\text{Log}(x + 24) - 2 \log 3 = \log(9-2x)$

CHAPTER TWENTY ONE: COMMERCIAL ARITHMETIC II

1. A business woman opened an account by depositing Kshs. 12,000 in a bank on 1st July 1995. Each subsequent year, she deposited the same amount on 1st July. The bank offered her 9% per annum compound interest. Calculate the total amount in her account on

    (a)   30th June 1996

    (b)   30th June 1997

2. A construction company requires to transport 144 tonnes of stones to sites A and B. The company pays Kshs 24,000 to transport 48 tonnes of stone for every 28 km. Kimani transported 96 tonnes to a site A, 49 km away.

    (a)   Find how much he paid

    (b)   Kimani spends Kshs 3,000 to transport every 8 tonnes of stones to site. Calculate his total profit.

    (c)   Achieng transported the remaining stones to sites B, 84 km away. If she made 44% profit, find her transport cost.

3. The table shows income tax rates

Monthly taxable pay	Rate of tax Kshs in 1 K£
1 – 435	2
436 – 870	3
871-1305	4
1306 – 1740	5
Excess Over 1740	6

    A company employee earn a monthly basic salary of Kshs 30,000 and is also given taxable allowances amounting to Kshs 10, 480.

(a)   Calculate the total income tax

(b)   The employee is entitled to a personal tax relief of Kshs 800 per month. Determine the net tax.

(c)   If the employee received a 50% increase in his total income, calculate the corresponding percentage increase on the income tax.

4. A house is to be sold either on cash basis or through a loan. The cash price is Kshs.750, 000. The loan conditions area as follows: there is to be down payment of 10% of the cash price and the rest of the money is to be paid through a loan at 10% per annum compound interest.

A customer decided to buy the house through a loan.

a)   (i)   Calculate the amount of money loaned to the customer.

        (ii)   The customer paid the loan in 3 year's. Calculate the total amount paid for the house.

b)   Find how long the customer would have taken to fully pay for the house if she paid a total of Kshs 891,750.

5. A businessman obtained a loan of Kshs. 450,000 from a bank to buy a matatu valued at the same amount. The bank charges interest at 24% per annum compound quarterly

a)   Calculate the total amount of money the businessman paid to clear the loan in 1 ½ years.

b)   The average income realized from the matatu per day was Kshs. 1500. The matatu worked for 3 years at an average of 280 days year. Calculate the total income from the matatu.

c) During the three years, the value of the matatu depreciated at the rate of 16% per annum. If the businessman sold the matatu at its new value, calculate the total profit he realized by the end of three years.

6. A bank either pays simple interest as 5% p.a or compound interest 5% p.a on deposits. Nekesa deposited Kshs P in the bank for two years on simple interest terms. If she had deposited the same amount for two years on compound interest terms, she would have earned Kshs 210 more.

Calculate without using Mathematics Tables, the values of P

7. (a) A certain sum of money is deposited in a bank that pays simple interest at a certain rate. After 5 years the total amount of money in an account is Kshs 358 400. The interest earned each year is 12 800

Calculate

(i) The amount of money which was deposited

(ii) The annual rate of interest that the bank paid

(b) A computer whose marked price is Kshs 40,000 is sold at Kshs 56,000 on hire purchase terms.

(i) Kioko bought the computer on hire purchase term. He paid a deposit of 25% of the hire purchase price and cleared the balance by equal monthly installments of Kshs 2625. Calculate the number of installments

(ii) Had Kioko bought the computer on cash terms he would have been allowed a discount of 12 ½ % on marked price. Calculate the difference between the cash price and the hire purchase price and express as a percentage of the cash price

(iii) Calculate the difference between the cash price and hire purchase price and express it as a percentage of the cash price.

8. The table below is a part of tax table for monthly income for the year 2004

Monthly taxable income In (Kshs)	Tax rate percentage (%) in each shillings
Under Kshs 9681	10%
From Kshs 9681 but under 18801	15%
From Kshs 18801 but 27921	20%

In the tax year 2004, the tax of Kerubo's monthly income was Kshs 1916.

Calculate Kerubo's monthly income

9. The cash price of a T.V set is Kshs 13, 800. A customer opts to buy the set on hire purchase terms by paying a deposit of Kshs 2280.

If simple interest of 20 p. a is charged on the balance and the customer is required to repay by 24 equal monthly installments. Calculate the amount of each installment.

10. A plot of land valued at Kshs. 50,000 at the start of 1994.

Thereafter, every year, it appreciated by 10% of its previous years value find:

(a) The value of the land at the start of 1995

(b) The value of the land at the end of 1997

11.   The table below shows Kenya tax rates in a certain year.

Income K £ per annum	Tax rates Kshs per K £
1- 4512	2
4513 – 9024	3
9025 – 13536	4
13537 – 18048	5
18049 – 22560	6
Over 22560	6.5

In that year Muhando earned a salary of Kshs. 16510 per month. He was entitled to a monthly tax relief of Kshs. 960

Calculate

(a)   Muhando's annual salary in K £

(b)   (i)   The monthly tax paid by Muhando in Kshs

14.   A tailor intends to buy a sewing machine which costs Kshs 48,000. He borrows the money from a bank. The loan has to be repaid at the end of the second year. The bank charges an interest at the rate of 24% per annum compounded half yearly. Calculate the total amount payable to the bank.

15.   The average rate of depreciation in value of a water pump is 9% per annum. After three complete years its value was Kshs 150,700. Find its value at the start of the three year period.

16.   A water pump costs Kshs 21600 when new, at the end of the first year its value

depreciates by 25%. The depreciation at the end of the second year is 20% and thereafter

the rate of depreciation is 15% yearly. Calculate the exact value of the water pump at the

end of the fourth year.

CHAPTER TWENTY TWO: CIRCLES, CHORDS AND TANGENTS

1. The figure below represents a circle a diameter 28 cm with a sector subtending an angle of 75^0 at the centre.

    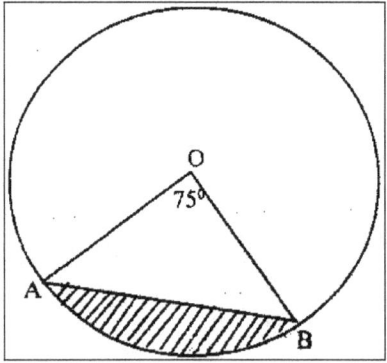

    Find the area of the shaded segment to 4 significant figures

    (a) <PST

2. The figure below represents a rectangle PQRS inscribed in a circle centre 0 and radius 17 cm. PQ = 16 cm.

    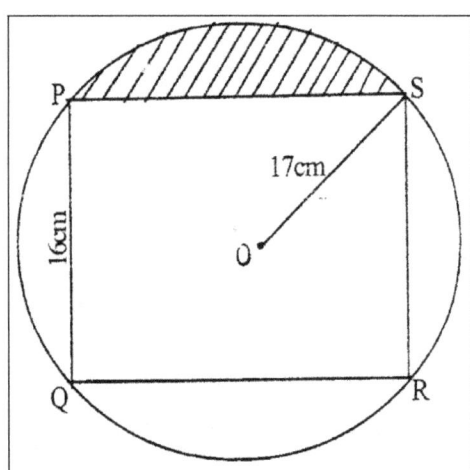

    Calculate

    (a) The length PS of the rectangle

    (b) The angle POS

    (c) The area of the shaded region

3. In the figure below, BT is a tangent to the circle at B. AXCT and BXD are straight lines. AX = 6 cm, CT = 8 cm, BX = 4.8 cm and XD = 5 cm.

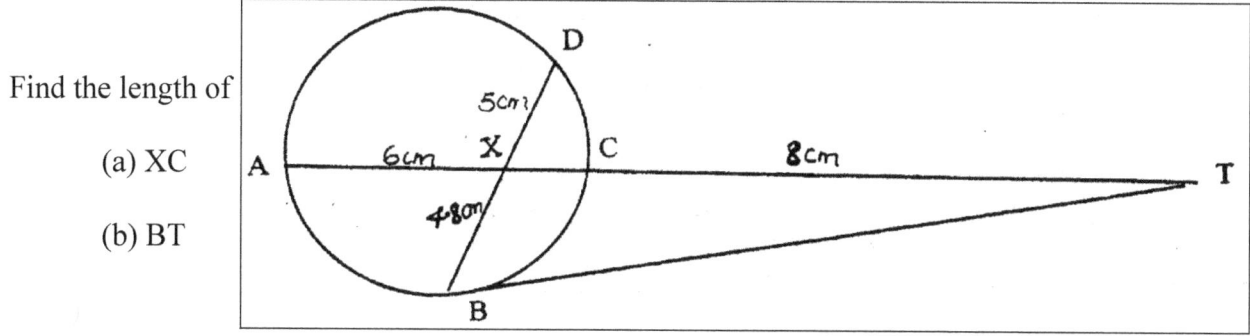

Find the length of

(a) XC

(b) BT

4.   The figure below shows two circles each of radius 7 cm, with centers at X and Y. The

circles touch each other at point Q.

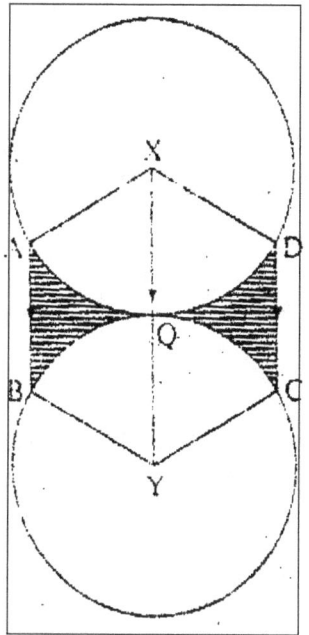

Given that <AXD = <BYC = 120⁰ and lines AB, XQY and DC are parallel, calculate the area of:

a)   Minor sector XAQD (Take $\pi \frac{22}{7}$)

b)   The trapezium XABY

c)   The shaded regions.

5.   The figure below shows a circle, centre, O of radius 7 cm. TP and TQ are tangents to the
     circle at points P and Q respectively. OT =25 cm.

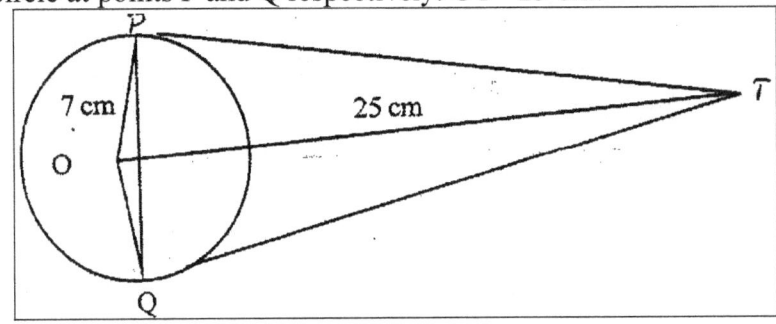

Calculate the length of the chord PQ

6.   The figure below shows a circle centre O and a point Q which is outside the circle

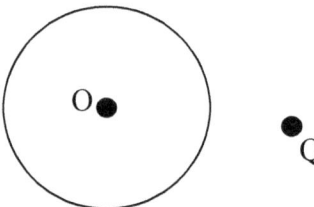

     Using a ruler and a pair of compasses, only locate a point on the circle such that angle

     OPQ = 90°

7.   In the figure below, PQR is an equilateral triangle of side 6 cm. Arcs QR, PR and PQ arcs
     of circles with centers at P, Q and R respectively.

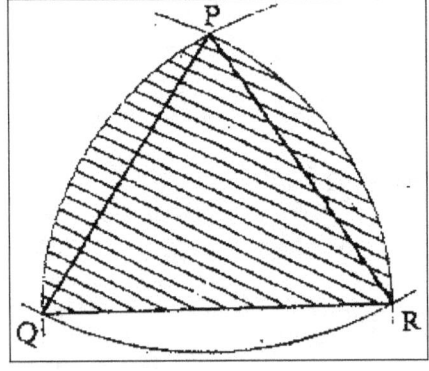

Calculate the area of the shaded region to 4 significant figures

8. In the figure below AB is a diameter of the circle. Chord PQ intersects AB at N. A tangent to the circle at B meets PQ produced at R.

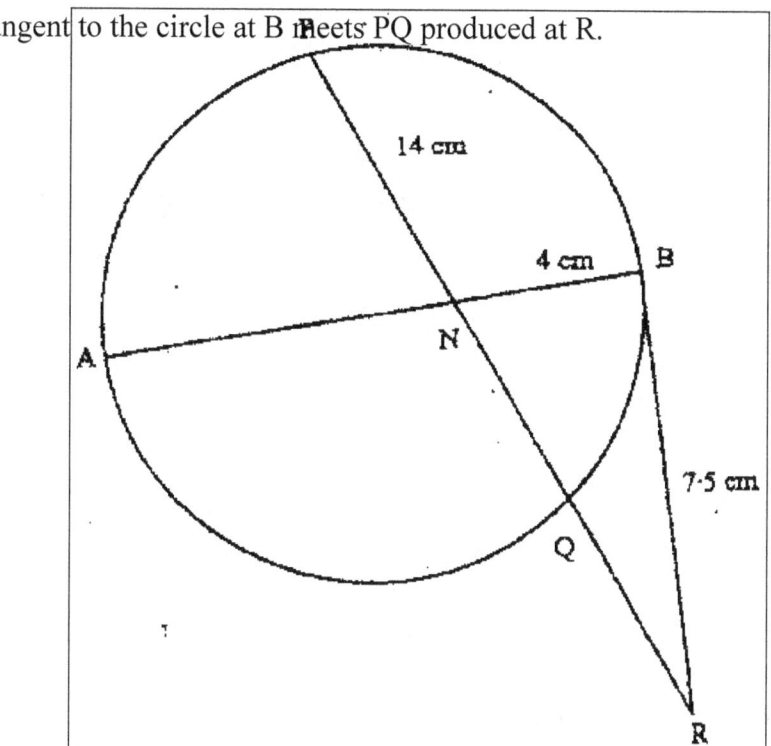

Given that PN = 14 cm, NB = 4 cm and BR = 7.5 cm, calculate the length of:

(a) NR

(b) AN

CHAPTER TWENTY THRE: MATRICES

1. A and B are two matrices. If A = $\begin{array}{cc} 1 & 2 \\ 4 & 3 \end{array}$ find B given that $A^2 = A + B$

2. Given that A= $\begin{array}{cc} 1 & 3 \\ 5 & 3 \end{array}$, B= $\begin{array}{cc} 3 & 1 \\ 5 & -1 \end{array}$, C = $\begin{array}{cc} p & 0 \\ 0 & q \end{array}$ and AB =BC, determine the value of P

3. A matrix A is given by A = $\begin{array}{cc} x & 0 \\ 5 & y \end{array}$

   a) Determine A^2

   b) If $A^2 = \begin{pmatrix} 1 & 0 \\ 0 & 1 \end{pmatrix}$, determine the possible pairs of values of x and y

4. (a) Find the inverse of the matrix $\begin{bmatrix} 9 & 8 \\ 7 & 6 \end{bmatrix}$

   (b) In a certain week a businessman bought 36 bicycles and 32 radios for total of Kshs 227 280. In the following week, he bought 28 bicycles and 24 radios for a total of Kshs 174 960. Using matrix method, find the price of each bicycle and each radio that he bought

   (c) In the third week, the price of each bicycle was reduced by 10% while the price of each radio was raised by 10%. The businessman bought as many bicycles and as many radios as he had bought in the first two weeks.
   Find by matrix method, the total cost of the bicycles and radios that the businessman bought in the third week.

5. Determine the inverse T^{-1} of the matrix $\begin{pmatrix} 1 & 2 \\ 1 & -1 \end{pmatrix}$

Hence find the coordinates to the point at which the two lines $x + 2y=7$ and $x-y=1$

6. Given that $A = \begin{pmatrix} 0 & -1 \\ 3 & 2 \end{pmatrix}$ and $B = \begin{pmatrix} -1 & 0 \\ 2 & -4 \end{pmatrix}$

Find the value of x if

(i) $A - 2x = 2B$

(ii) $3x - 2A = 3B$

(iii) $2A - 3B = 2x$

7. Find the non- zero value of k for which $\begin{pmatrix} k+1 & 2 \\ 4k & 2k \end{pmatrix}$ is an inverse.

8. A clothes dealer sold 3 shirts and 2 trousers for Kshs. 840 and 4 shirts and 5 trousers for Kshs 1680. Form a matrix equation to represent the above information. Hence find the cost of 1 shirt and the cost of 1 trouser.

CHAPTER TWENTY FOUR: FORMULAE AND VARIATIONS

1. The volume Vcm^3 of an object is given by

$$V = \frac{2}{3} \pi r^3 \left(\frac{1}{sc^2} - 2 \right)$$

   Express in term of π r, s and V

2. Make V the subject of the formula

$$T = \frac{1}{2} m (u^2 - v^2)$$

3. Given that $y = \dfrac{b - bx^2}{cx^2 - a}$ make x the subject

4. Given that log y = log (10^n) make n the subject

5. A quantity T is partly constant and partly varies as the square root of S.

   i. Using constants a and b, write down an equation connecting T and S.

   ii. If S = 16, when T = 24 and S = 36 when T = 32, find the values of the

   constants a and b,

6. A quantity P is partly constant and partly varies inversely as a quantity q, given that p = 10 when q = 1.5 and p = 20, when q = 1.25, find the value of p when q= 0.5

7. Make y the subject of the formula $p = \dfrac{xy}{x-y}$

8. Make P the subject of the formula

   $P^2 = (P - q) (P-r)$

9.	The density of a solid spherical ball varies directly as its mass and inversely as the cube of its radius

When the mass of the ball is 500g and the radius is 5 cm, its density is 2 g per cm^3

Calculate the radius of a solid spherical ball of mass 540 density of 10g per cm^3

10.	Make s the subject of the formula

$$\sqrt{P} = r\sqrt{1 - as^2}$$

11.	The quantities t, x and y are such that t varies directly as x and inversely as the square root of y. Find the percentage in t if x decreases by 4% when y increases by 44%

12.	Given that y is inversely proportional to x^n and k as the constant of proportionality;

	(a)	(i)	Write down a formula connecting y, x, n and k

		(ii)	If x = 2 when y = 12 and x = 4 when y = 3, write down two expressions for k in terms of n.

			Hence, find the value of n and k.

	(b)	Using the value of n obtained in (a) (ii) above, find y when x = 5 $\frac{1}{3}$

13.	The electrical resistance, R ohms of a wire of a given length is inversely proportional to the square of the diameter of the wire, d mm. If R = 2.0 ohms when d = 3mm. Find the vale R when d = 4 mm.

14.	The volume Vcm^3 of a solid depends partly on r and partly on r where rcm is one of the dimensions of the solid.

	When r = 1, the volume is 54.6 cm^3 and when r = 2, the volume is 226.8 cm^3

	(a) Find an expression for V in terms of r

(b) Calculate the volume of the solid when r = 4

(c) Find the value of r for which the two parts of the volume are equal

15. The mass of a certain metal rod varies jointly as its length and the square of its radius. A rod 40 cm long and radius 5 cm has a mass of 6 kg. Find the mass of a similar rod of length 25 cm and radius 8 cm.

16. Make x the subject of the formula

$$P = \frac{xy}{z + x}$$

17. The charge c shillings per person for a certain service is partly fixed and partly inversely proportional to the total number N of people.

(a) Write an expression for c in terms on N

(b) When 100 people attended the charge is Kshs 8700 per person while for 35 people the charge is Kshs 10000 per person.

(c) If a person had paid the full amount charge is refunded. A group of people paid but ten percent of organizer remained with Kshs 574000.
Find the number of people.

18. Two variables A and B are such that A varies partly as B and partly as the square root of B given that A=30, when B=9 and A=16 when B=14, find A when B=36.

19. Make p the subject of the formula

$$A = \frac{-EP}{\sqrt{P^2 + N}}$$

CHAPTER TWENTY FIVE: SEQUENCE AND SERIES

1.   The first, the third and the seventh terms of an increasing arithmetic progression are three consecutive terms of a geometric progression. In the first term of the arithmetic progression is 10 find the common difference of the arithmetic progression.

2.   Kubai saved Kshs 2,000 during the first year of employment. In each subsequent year, he saved 15% more than the preceding year until he retired.

     (a) How much did he save in the second year?

     (b) How much did he save in the third year?

     (c) Find the common ratio between the savings in two consecutive years

     (d) How many years did he take to save the savings a sum of Kshs 58,000?

     (e) How much had he saved after 20 years of service?

3.   In geometric progression, the first is a and the common ratio is r. The sum of the first two terms is 12 and the third term is 16.

     (a) Determine the ratio $\underline{ar^2}$

$$a + ar$$

     (b) If the first term is larger than the second term, find the value of r.

4.   (a)   The first term of an arithmetic progression is 4 and the last term is 20. The sum of the term is 252. Calculate the number of terms and the common differences of the arithmetic progression

     (b)   An Experimental culture has an initial population of 50 bacteria. The population increased by 80% every 20 minutes. Determine the time it will take to have a population of 1.2 million bacteria.

5. Each month, for 40 months, Amina deposited some money in a saving scheme. In the first month she deposited Kshs 500. Thereafter she increased her deposits by Kshs. 50 every month.

Calculate the:

a) Last amount deposited by Amina

b) Total amount Amina had saved in the 40 months.

6. A carpenter wishes to make a ladder with 15 cross- pieces. The cross- pieces are to diminish uniformly in length from 67 cm at the bottom to 32 cm at the top.

Calculate the length in cm, of the seventh cross- piece from the bottom

7. The second and fifth terms of a geometric progression are 16 and 2 respectively. Determine the common ratio and the first term.

8. The eleventh term of an arithmetic progression is four times its second term. The sum of the first seven terms of the same progression is 175

(a) Find the first term and common difference of the progression

(b) Given that p^{th} term of the progression is greater than 124, find the least value of P

9. The nth term of sequence is given by $2n + 3$ of the sequence

(a) Write down the first four terms of the sequence

(b) Find s_n the sum of the fifty term of the sequence

(c) Show that the sum of the first n terms of the sequence is given by

$S_n = n^2 + 4n$

Hence or otherwise find the largest integral value of n such that Sn <725

CHAPTER TWENTY SIX: VECTORS

1.   The figure below is a right pyramid with a rectangular base ABCD and VO as the height.

The vectors AD= a, AB = b and DV = v

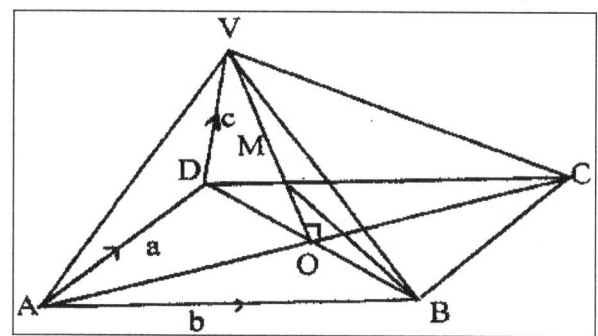

a) Express

   (i)   AV in terms of a and c

   (ii)   BV in terms of a, b and c

(b) M is point on OV such that OM: MV=3:4, Express BM in terms of a, b and c.

Simplify your answer as far as possible

2.   In triangle OAB, OA = a OB = b and P lies on AB such that AP: BP = 3.5

(a) Find the terms of a and b the vectors

   (i)   AB

   (ii)   AP

   (iii)   BP

   (iv)   OP

(b) Point Q is on OP such AQ = $\frac{-5 + 9}{8a \quad 40b}$

Find the ratio OQ: QP

3. The figure below shows triangle OAB in which M divides OA in the ratio 2: 3 and N

    divides OB in the ratio 4:1 AN and BM intersect at X

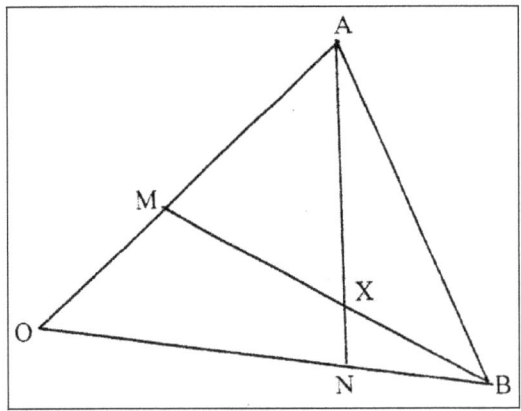

(a) Given that OA = a and OB = b, express in terms of a and b:

(i)   AN

(ii)   BM

(b) If AX = s AN and BX = tBM, where s and t are constants, write two expressions

for OX in terms of a,b s and t

Find the value of s

Hence write OX in terms of a and b

4. The position vectors for points P and Q are 4 I + 3 j + 6 j + 6 k respectively. Express

vector PQ in terms of unit vectors I, j and k. Hence find the length of PQ, leaving your

answer in simplified surd form.

5. In the figure below, vector OP = P and OR =r. Vector OS = 2r and OQ = $^3/_2$p.

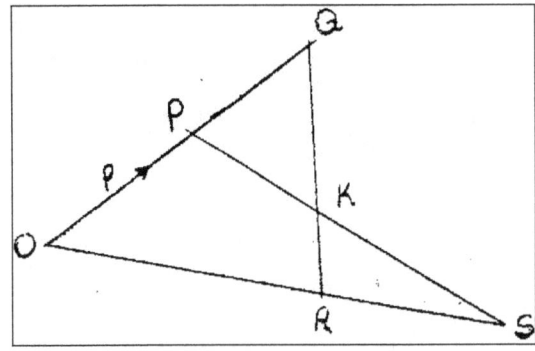

102

a)   Express in terms of p and r (i) QR and (ii) PS

    b)   The lines QR and PS intersect at K such that QK = m QR and PK = n PS, where

        m and n are scalars. Find two distinct expressions for OK in terms of p,r,m and n.

        Hence find the values of m and n.

    c)   State the ratio PK: KS

6.   Point T is the midpoint of a straight line AB. Given the position vectors of A and T are i-j

+ k and 2i+ 1½ k respectively, find the position vector of B in terms of i, j and k

7.   A point R divides a line PQ internally in the ration 3:4. Another point S, divides the line

PR externally in the ration 5:2. Given that PQ = 8 cm, calculate the length of RS, correct

to 2 decimal places.

8.   The points P, Q, R and S have position vectors 2p, 3p, r and 3r respectively, relative to an

origin O. A point T divides PS internally in the ratio 1:6

    (a)   Find, in the simplest form, the vectors OT and QT in terms p and r

    (b)   (i)   Show that the points Q, T, and R lie on a straight line

           (ii)   Determine the ratio in which T divides QR

9.   Two points P and Q have coordinates (-2, 3) and (1, 3) respectively. A translation map

point P to P' (10, 10)

(a) Find the coordinates of Q' the image of Q under the translation

(b) The position vector of P and Q in (a) above are p and q respectively given that mp –

nq = -12 $\begin{pmatrix} \\ 9 \end{pmatrix}$

    Find the value of m and n

10.   Given that q i + $\frac{1}{3}$ j + $\frac{2}{3}$ k is a unit vector, find q

11. In the diagram below, the coordinates of points A and B are (1, 6) and (15, 6) respectively). Point N is on OB such that 3 ON = 2 OB. Line OA is produced to L such that OL = 3 OA

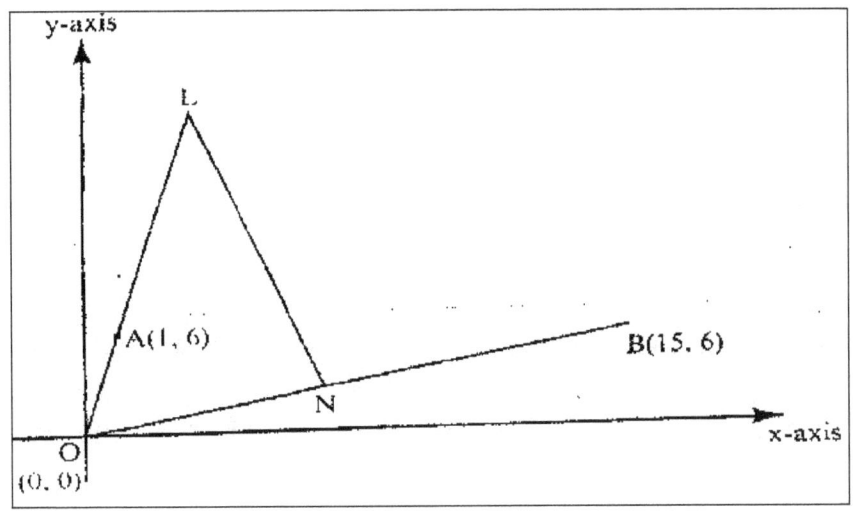

(a) Find vector LN

(b) Given that a point M is on LN such that LM: MN = 3: 4, find the coordinates of

M

(c) If line OM is produced to T such that OM: MT = 6:1

(i)   Find the position vector of T

(ii)   Show that points L, T and B are collinear

12. In the figure below, OQ = q and OR = r. Point X divides OQ in the ratio 1: 2 and Y divides OR in the ratio 3: 4 lines XR and YQ intersect at E.

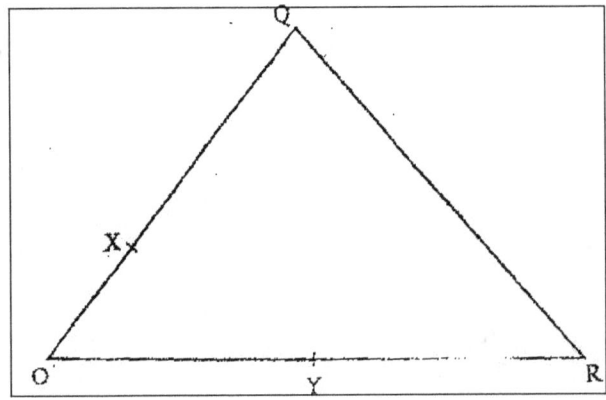

(a) Express in terms of q and r

    (i)   XR

    (ii)   YQ

(b) If XE = m XR and YE = n YQ, express OE in terms of:

    (i)   r, q and m

    (ii)   r, q and n

(c) Using the results in (b) above, find the values of m and n.

13.   Vector q has a magnitude of 7 and is parallel to vector p. Given that

    p= 3 i –j + 1 ½ k, express vector q in terms of i, j, and k.

14.   In the figure below, OA = 3i + 3j ABD OB = 8i – j. C is a point on AB such that AC:CB

    3:2, and D is a point such that OB//CD and 2OB = CD (T17)

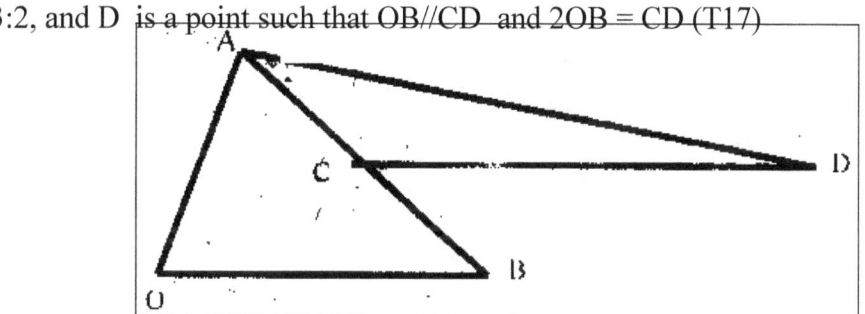

Determine the vector DA in terms of I and j

15.   In the figure below, KLMN is a trapezium in which KL is parallel to NM and KL = 3NM

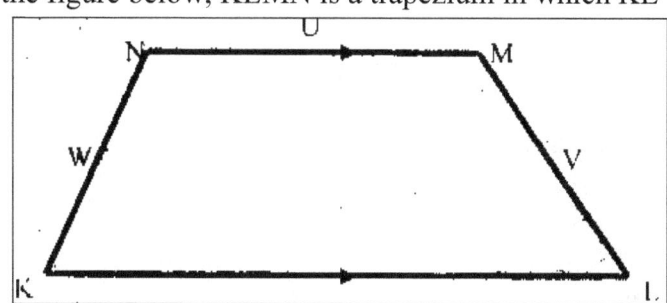

Given that KN = w, NM = u and ML = v. Show that 2u = v + w

16. The points P, Q and R lie on a straight line. The position vectors of P and R are 2i + 3j + 13k and 5i – 3j + 4k respectively; Q divides SR internally in the ratio 2: 1. Find the

(a) Position vector of Q

(b) Distance of Q from the origin

17. Co-ordinates of points O, P, Q and R are (0, 0), (3, 4), (11, 6) and (8, 2) respectively. A point T is such that the vector OT, QP and QR satisfy the vector equation OT = QP ½ QT. Find the coordinates of T.

18. In the figure below OA = a, OB = b, AB = BC and OB: BD = 3:1

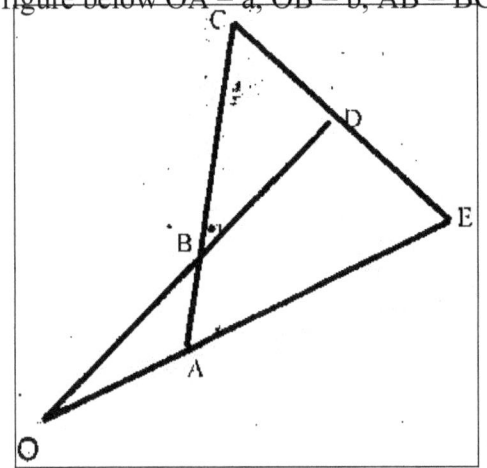

(a) Determine

  (i) AB in terms of a and b

  (ii) CD, in terms of a and b

(b) If CD: DE = 1 k and OA: AE = 1m determine

  (i) DE in terms of a, b and k

  (ii) The values of k and m

19. The figure below shows a grid of equally spaced parallel lines

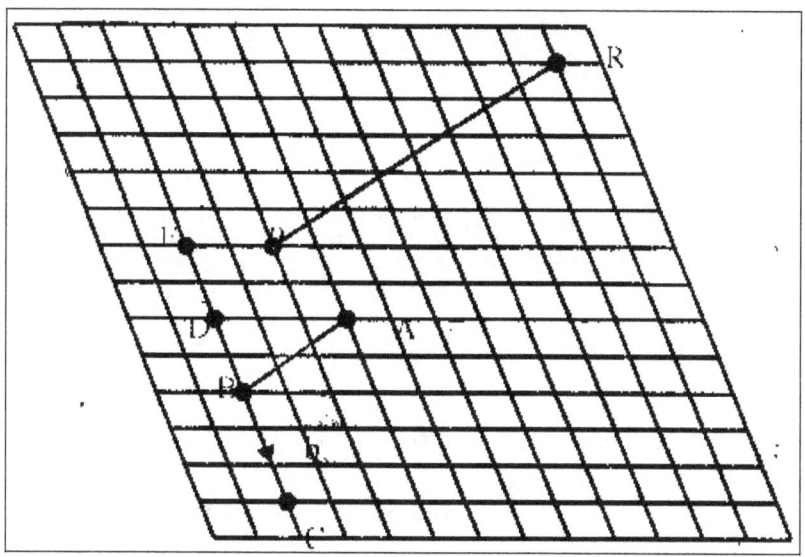

$\overrightarrow{AB}$ = a and $\overrightarrow{BC}$ = b

(a)   Express

    (i) $\overrightarrow{AC}$ in terms of a and b

    (ii)   AD in terms of a and b

(b)   Using triangle BEP, express BP in terms of a and b

(c)   PR produced meets BA produced at X and PR = $\frac{1}{9}$b – $\frac{8}{3}$a

By writing PX as kPR and BX as hBA and using the triangle BPX determine the

ratio PR: RX

20.   The position vectors of points x and y are x = 2i + j – 3k and y = 3i + 2j – 2k respectively.

Find XY

21.   Given that X = 2i + j -2K, y = -3i + 4j – k and z= 5i + 3j + 2k and that p= 3x – y + 2z,

find the magnitude of vector p to 3 significant figures.

CHAPTER TWENTY SEVEN: BINOMIAL

EXPRESSION

1.   (a)   Write down the simplest expansion $(1 + x)^6$

        (b)   Use the expansion up to the fourth term to find the value of $(1.03)^6$ to the nearest one thousandth.

2.   Use binomial expression to evaluate $(0.96)^5$ correct to 4 significant figures.

3.   Expand and simplify $(3x - y)^4$ hence use the first three terms of the expansion to proximate the value of $(6 - 0.2)^4$

4.   Abdi and Amoit were employed at the begging of the same year. Their annual salaries in shillings progressed as follows

    Abdi: 60000, 64800, 69600

    Amoit: 60000, 64800, 69984

        (a)   Calculate Abdi's annual salary increment and hence write down an expression for his annual salary in his n^{th} year of employment?

        (b)   Calculate Amoit's annual percentage rate of salary increment and hence write down an expression for her annual salary in her n^{th} year employment?

        (c)   Calculate the difference in the annual salary for Abdi and Amoit in their 7th year of employment.

5.   Use binomial expression to evaluate

$$\left(2 + \frac{1}{\sqrt{2}}\right)^5 + \left(2 - \frac{1}{\sqrt{2}}\right)^5$$

6.   (a)   Expand the expression $\left(1 + \underline{1}x\right)^5$ in ascending powers of x, leaving

2

the coefficients as fractions in their simplest form.

(b)   Use the first three terms of the expression in (a) above to estimate the value of

$$\left(1\frac{1}{20}\right)^5$$

7.   (a)   Expand $(a - b)^6$

(b)   Use the first three terms of the expansion in (a) above to find the approximate

value of $(1.98)^6$

8.   Expand $(2 + x)^5$ in ascending powers of x up to the term in x^3 hence approximate the

value of $(2.03)^5$ to 4 s.f

9.   (a)   Expand $(1 + x)^5$

Hence use the expansion to estimate $(1.04)^5$ correct to 4 decimal places

(b)   Use the expansion up to the fourth term to find the value of $(1.03)^6$ to the nearest

one thousandth.

10.   Expand and Simplify $(1-3x)^5$ up to the term in x^3

Hence use your expansion to estimate $(0.97)^5$ correct to decimal places.

11.   Expand $(1 + a)^5$

Use your expansion to evaluate $(0.8)^5$ correct to four places of decimal

12.   (a)   Expand $(1 + x)^5$

(b)   Use the first three terms of the expansion in (a) above to find the approximate value of

$(0.98)^5$

CHAPTER TWENTY EIGHT: PROBABILITY

1. The probabilities that a husband and wife will be alive 25 years from now are 0.7 and 0.9 respectively.

    Find the probability that in 25 years time,

    (a) Both will be alive

    (b) Neither will be alive

    (c) One will be alive

    (d) At least one will be alive

2. A bag contains blue, green and red pens of the same type in the ratio 8:2:5 respectively.

    A pen is picked at random without replacement and its colour noted

    (a)   Determine the probability that the first pen picked is

        (i)   Blue

        (ii)   Either green or red

    (b)   Using a tree diagram, determine the probability that

        (i)   The first two pens picked are both green

        (ii)   Only one of the first two pens picked is red.

3. A science club is made up of boys and girls. The club has 3 officials. Using a tree diagram or otherwise find the probability that:

    (a) The club officials are all boys

    (b) Two of the officials are girls

4. Two baskets A and B each contain a mixture of oranges and limes, all of the same size. Basket A contains 26 oranges and 13 limes. Basket B contains 18 oranges and 15 limes. A child selected a basket at random and picked a fruit at a random from it.

(a) Illustrate this information by a probabilities tree diagram

(b) Find the probability that the fruit picked was an orange.

5. In form 1 class there are 22 girls and boys. The probability of a girl completing the secondary education course is 3 whereas that of a boy is $^2/_3$

   (a) A student is picked at random from class. Find the possibility that,

   (i) The student picked is a boy and will complete the course

   (ii) The student picked will complete the course

   (b) Two students are picked at random. Find the possibility that they are a boy and a girl and that both will not complete the course.

6. Three representatives are to be selected randomly from a group of 7 girls and 8 boys. Calculate the probability of selecting two girls and one boy.

7. A poultry farmer vaccinated 540 of his 720 chickens against a disease. Two months later, 5% of the vaccinated and 80% of the unvaccinated chicken, contracted the disease. Calculate the probability that a chicken chosen random contacted the disease.

8. The probability of three darts players Akinyi, Kamau, and Juma hitting the bulls eye are 0.2, 0.3 and 1.5 respectively.

   (a) Draw a probability tree diagram to show the possible outcomes

   (b) Find the probability that:

   (i) All hit the bull's eye

   (ii) Only one of them hit the bull's eye

   (iii) At most one missed the bull's eye

9. (a) An unbiased coin with two faces, head (H) and tail (T), is tossed three times, list all the possible outcomes.

Hence determine the probability of getting:

(i)   At least two heads

(ii)   Only one tail

(b)   During a certain motor rally it is predicted that the weather will be either dry (D) or wet (W). The probability that the weather will be dry is estimated to be $\frac{7}{10}$. The probability for a driver to complete (C) the rally during the dry weather is estimated to be $\frac{5}{6}$. The probability for a driver to complete the rally during wet weather is estimated to be $\frac{1}{10}$. Complete the probability tree diagram given below.

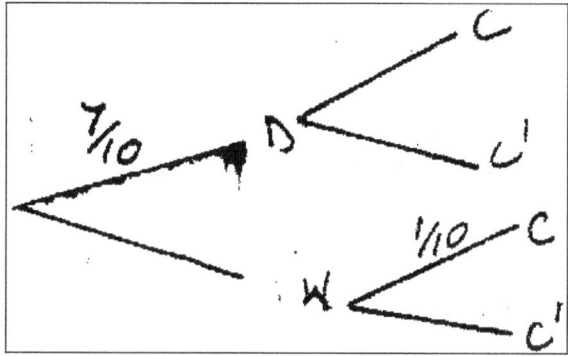

What is the probability that:

(i)   The driver completes the rally?

(ii)   The weather was wet and the driver did not complete the rally?

10.   There are three cars A, B and C in a race. A is twice as likely to win as B while B is twice as likely to win as c. Find the probability that.

a)   A wins the race

b)   Either B or C wins the race.

11. In the year 2003, the population of a certain district was 1.8 million. Thirty per cent of the population was in the age group 15 – 40 years. In the same year, 120,000 people in the district visited the Voluntary Counseling and Testing (VCT) centre for an HIV test. If a person was selected at random from the district in this year. Find the probability that the person visited a VCT centre and was in the age group 15 – 40 years.

12. (a) Two integers x and y are selected at random from the integers 1 to 8. If the same integer may be selected twice, find the probability that

   (i) $|x - y| = 2$

   (ii) $|x - y|$ is 5 or more

   (iii) $x > y$

   (b) A die is biased so that when tossed, the probability of a number r showing up, is given by $p(r) = Kr$ where K is a constant and $r = 1, 2,3,4,5$ and 6 (the number on the faces of the die

   (i) Find the value of K

   (ii) If the die is tossed twice, calculate the probability that the total score is 11

13. Two bags A and B contain identical balls except for the colours. Bag A contains 4 red balls and 2 yellow balls. Bag B contains 2 red balls and 3 yellow balls.

   (a) If a ball is drawn at random from each bag, find the probability that both balls are of the same colour.

   (b) If two balls are drawn at random from each bag, one at a time without replacement, find the probability that:

   (i) The two balls drawn from bag A or bag B are red

(ii)   All the four balls drawn are red

14.   During inter – school competitions, football and volleyball teams from Mokagu high school took part. The probability that their football and volleyball teams would win were $^3/_8$ and $^4/_7$ respectively.

Find the probability that

(a)   Both their football and volleyball teams

(b)   At least one of their teams won

15.   A science club is made up of 5 boys and 7 girls. The club has 3 officials. Using a tree diagram or otherwise find the probability that:

(a)   The club officials are all boys

(b)   Two of the officials are girls

16.   Chicks on Onyango's farm were noted to have either brown feathers brown or black tail feathers. Of those with black feathers 2/3 were female while $^2/_5$ of those with brown feathers were male. Otieno bought two chicks from Onyango. One had black tail feathers while the other had brown find the probability that Otieno's chicks were not of the same gender

17.   Three representatives are to be selected randomly from a group of 7 girls and 8 boys. Calculate the probability of selecting two girls and one boy

18. The probability that a man wins a game is ¾. He plays the game until he wins. Determine the probability that he wins in the fifth round.

19. The probability that Kamau will be selected for his school's basketball team is ¼. If he is selected for the basketball team. Then the probability that he will be selected for football is $^{1}/_{3}$ if he is not selected for basketball then the probability that he is selected for football is $^{4}/_{5}$. What is the probability that Kamau is selected for at least one of the two games?

20. Two baskets A and B each contains a mixture of oranges and lemons. Baskets A contains 26 oranges and 13 lemons. Baskets B contains 18 oranges and 15 lemons. A child selected a basket at random and picked at random a fruit from it. Determine the probability that the fruit picked was an orange.

CHAPTER TWENTY NINE: COMPOUND PROPORTION

AND MIXTURES

1. Akinyi bought and beans from a wholesaler. She then mixed the maize and beans the ratio 4:3 she brought the maize as Kshs. 12 per kg and the beans 4 per kg. If she was to make a profit of 30% what should be the selling price of 1 kg of the mixture?

2. A rectangular tank of base 2.4 m by 2.8 m and a height of 3 m contains 3,600 liters of water initially. Water flows into the tank at the rate of 0.5 litres per second

   Calculate the time in hours and minutes, required to fill the tank

3. A company is to construct a parking bay whose area is 135m². It is to be covered with concrete slab of uniform thickness of 0.15. To make the slab cement. Ballast and sand are to be mixed so that their masses are in the ratio 1: 4: 4. The mass of m³ of dry slab is 2, 500kg.

   Calculate

   (a)   (i)   The volume of the slab

          (ii)   The mass of the dry slab

          (iii) The mass of cement to be used

   (b)   If one bag of the cement is 50 kg, find the number of bags to be purchased

   (c)   If a lorry carries 7 tonnes of sand, calculate the number of lorries of sand
          to be purchased.

4. The mass of a mixture A of beans and maize is 72 kg. The ratio of beans to maize is 3:5 respectively

   (a)   Find the mass of maize in the mixture

(b)   A second mixture of B of beans and maize of mass 98 kg in mixed with A. The final ratio of beans to maize is 8:9 respectively. Find the ratio of beans to maize in B

5.   A retailer bought 49 kg of grade 1 rice at Kshs. 65 per kilogram and 60 kg of grade II rice at Kshs 27.50 per kilogram. He mixed the tow types of rice.

(a) Find the buying price of one kilogram of the mixture

(b) He packed the mixture into 2 kg packets

(i)   If he intends to make a 20% profit find the selling price per packet

(ii)   He sold 8 packets and then reduced the price by 10% in order to attract customers. Find the new selling price per packet.

(iii)   After selling $\frac{1}{3}$ of the remainder at reduced price, he raised the price so as to realize the original goal of 20% profit overall. Find the selling price per packet of the remaining rice.

6.   A trader sells a bag of beans for Kshs 1,200. He mixed beans and maize in the ration 3: 2. Find how much the trader should he sell a bag of the mixture to realize the same profit?

7.   Pipe A can fill an empty water tank in 3 hours while, pipe B can fill the same tank in 6 hours, when the tank is full it can be emptied by pipe C in 8 hours. Pipes A and B are opened at the same time when the tank is empty.

If one hour later, pipe C is also opened, find the total time taken to fill the tank

8.   A solution whose volume is 80 litres is made 40% of water and 60% of alcohol. When litres of water are added, the percentage of alcohol drops to 40%

(a)   Find the value of x

(b)   Thirty litres of water is added to the new solution. Calculate the percentage

118

(c)   If 5 litres of the solution in (b) is added to 2 litres of the original solution, calculate in the simplest form, the ratio of water to that of alcohol in the resulting solution

9.   A tank has two inlet taps P and Q and an outlet tap R. when empty, the tank can be filled by tap P alone in 4 ½ hours or by tap Q alone in 3 hours. When full, the tank can be emptied in 2 hours by tap R.

(a)   The tank is initially empty. Find how long it would take to fill up the tank

(i)   If tap R is closed and taps P and Q are opened at the same time

(ii)   If all the three taps are opened at the same time

(b)   The tank is initially empty and the three taps are opened as follows

P at 8.00 a.m

Q at 8.45 a.m

R at 9.00 a.m

(i)   Find the fraction of the tank that would be filled by 9.00 a.m

(ii)   Find the time the tank would be fully filled up

10.   Kipketer can cultivate a piece of land in 7 hrs while Wanjiru can do the same work in 5 hours. Find the time they would take to cultivate the piece of land when working together.

11. Mogaka and Ondiso working together can do a piece of work in 6 days. Mogaka, working alone, takes 5 days longer than Onduso. How many days does it take Onduso to do the work alone.

12. Wainaina has two dairy farms A and B. Farm A produces milk with 3 ¼ percent fat and farm B produces milk with 4 ¼ percent fat.

   (a) (i) The total mass of milk fat in 50 kg of milk from farm A and 30kg of milk from farm B.

       (ii) The percentage of fat in a mixture of 50 kg of milk A and 30 kg of milk from B

   (b) Determine the range of values of mass of milk from farm B that must be used in a 50 kg mixture so that the mixture may have at least 4 percent fat.

13. A construction firm has two tractors T_1 and T_2. Both tractors working together can complete the work in 6 days while T_1 alone can complete the work in 15 days. After the two tractors had worked together for four days, tractor T1 broke down.
    Find the time taken by tractor T_2 complete the remaining work.

14. The points P, Q, R and S have position vectors 2p, 3 p, r and 3r respectively, relative to an origin O. A point T divides PS internally in the ratio 1: 6

   (a) Find in the simplest form, the vectors OT and QT in terms of P and r

   (b) (i) Show that the points Q, T and R lie on a straight line.

       (ii) Determine the ratio in which T divides QR.

CHAPTER THIRTY: GRAPHICAL METHODS

1. The table shows the height metres of an object thrown vertically upwards varies with the time t seconds

   The relationship between s and t is represented by the equations $s = at^2 + bt + 10$ where b are constants.

t	0	1	2	3	4	5	6	7	8	9	10
s		45.1						49.9			-80

   (c) (i) Using the information in the table, determine the values of a and b

       (ii) Complete the table

   (b) (i) Draw a graph to represent the relationship between s and t

       (ii) Using the graph determine the velocity of the object when t = 5 seconds

2. Data collected form an experiment involving two variables X and Y was recorded as shown in the table below

x	1.1	1.2	1.3	1.4	1.5	1.6
y	-0.3	0.5	1.4	2.5	3.8	5.2

   The variables are known to satisfy a relation of the form $y = ax^3 + b$ where a and b are constants

   (a) For each value of x in the table above, write down the value of x^3

   (b)   (i) By drawing a suitable straight line graph, estimate the values of a and b

       (ii) Write down the relationship connecting y and x

3. Two quantities P and r are connected by the equation $p = kr^n$. The table of values of P and r is given below.

P	1.2	1.5	2.0	2.5	3.5	4.5
R	1.58	2.25	3.39	4.74	7.86	11.5

a)   State a liner equation connecting P and r.

b)   Using the scale 2 cm to represent 0.1 units on both axes, draw a suitable

       line graph on the grid provided. Hence estimate the values of K and n.

4.   The points which coordinates (5,5) and (-3,-1) are the ends of a diameter of a circle

centre A

Determine:

(a)   The coordinates of A

       The equation of the circle, expressing it in form $x^2 + y^2 + ax + by + c = 0$

       where a, b, and c are constants each computer sold

5.   The figure below is a sketch of the graph of the quadratic function $y = k$

       (x+1) (x-2)

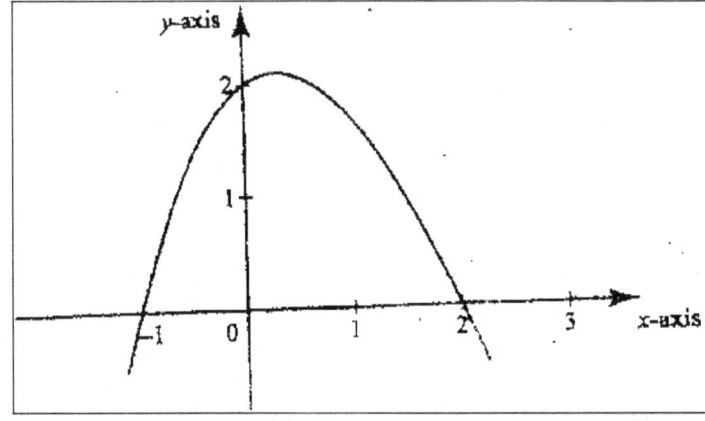

Find the value of k

6.   The table below shows the values of the length X (in metres) of a pendulum and the

corresponding values of the period T (in seconds) of its oscillations obtained in an experiment.

122

X (metres)	0.4	1.0	1.2	1.4	1.6
T (seconds)	1.25	2.01	2.19	2.37	2.53

(a) Construct a table of values of log X and corresponding values of log T, correcting each

value to 2 decimal places

(b) Given that the relation between the values of log X and log T approximate to a

linear law of the form m log X + log a where a and b are constants

(i) Use the axes on the grid provided to draw the line of best fit for the

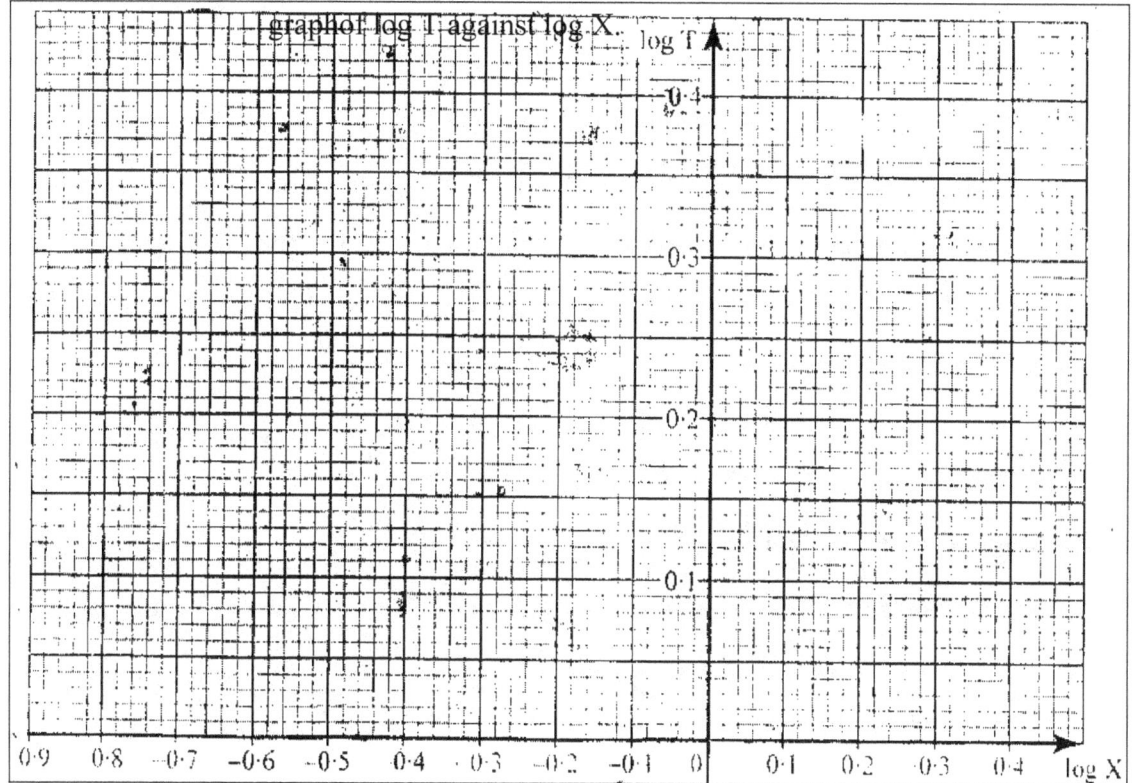

(ii) Use the graph to estimate the values of a and b

(iii) Find, to decimal places the length of the pendulum whose period is 1

second.

123

7. Data collection from an experiment involving two variables x and y was recorded as shown in the table below

X	1.1	1.2	1.3	1.4	1.5	1.6
Y	-0.3	0.5	1.4	2.5	3.8	5.2

The variables are known to satisfy a relation of the form $y = ax^3 + b$ where a and b are constants

(a) For each value of x in the table above. Write down the value of x^3

(b) (i) By drawing s suitable straight line graph, estimate the values of a and b

(ii) Write down the relationship connecting y and x

8. Two variables x and y, are linked by the relation $y = ax^n$. The figure below shows part of the straight line graph obtained when log y is plotted against log x.

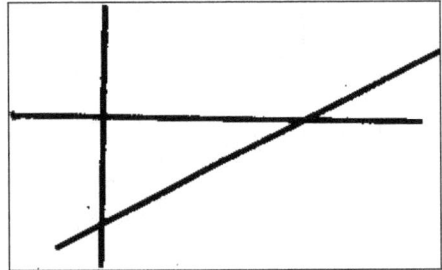

Calculate the value of a and n

9. The luminous intensity I of a lamp was measured for various values of voltage v across it.

The results were as shown below

V(volts)	30	36	40	44	48	50	54
L (Lux)	708	1248	1726	2320	3038	3848	4380

It is believed that V and l are related by an equation of the form $l = aV^n$ where a and n are constant.

(a)   Draw a suitable linear graph and determine the values of a and n

(b)   From the graph find

      (i)   The value of I when V = 52

      (ii)   The value of V when I = 2800

10.   In a certain relation, the value of A and B observe a relation $B = CA + KA^2$ where C and K are constants. Below is a table of values of A and B

A	1	2	3	4	5	6
B	3.2	6.75	10.8	15.1	20	25.2

(a) By drawing a suitable straight line graphs, determine the values of C and K.

(b) Hence write down the relationship between A and B

(c) Determine the value of B when A = 7

11.   The variables P and Q are connected by the equation $P = ab^q$ where a and b are constants. The value of p and q are given below

P	6.56	17.7	47.8	129	349	941	2540	6860
q	0	1	2	3	4	5	6	7

(a)   State the equation in terms of p and q which gives a straight line graph

(b)   By drawing a straight line graph, estimate the value of constants a and b and give

your answer correct to 1 decimal place.

CHAPTER THIRTY ONE: MATRICES AND TRANSFORMATIONS

1. Matrix p is given by $\begin{pmatrix} 1 & 2 \\ 4 & 3 \end{pmatrix}$

   (a) Find P^{-1}

   (b) Two institutions, Elimu and Somo, purchase beans at Kshs. B per bag and maize at Kshs m per bag. Elimu purchased 8 bags of beans and 14 bags of maize for Kshs 47,600. Somo purchased 10 bags of beans and 16 of maize for Kshs. 57,400

   (c) The price of beans later went up by 5% and that of maize remained constant. Elimu bought the same quantity of beans but spent the same total amount of money as before on the two items. State the new ratio of beans to maize.

2. A triangle is formed by the coordinates A (2, 1) B (4, 1) and C (1, 6). It is rotated clockwise through 90^0 about the origin. Find the coordinates of this image.

3. On the grid provided on the opposite page A (1, 2) B (7, 2) C (4, 4) D (3, 4) is a trapezium

(a) ABCD is mapped onto A'B'C'D' by a positive quarter turn. Draw the image A'B'C'D on the grid

(b) A transformation $\begin{pmatrix} -2 & -1 \\ 0 & 1 \end{pmatrix}$ maps A'B'C'D onto A"B" C"D" Find the coordinates of A"B"C"D"

4. A triangle T whose vertices are A (2, 3) B (5, 3) and C (4, 1) is mapped onto triangle T^1 whose vertices are A^1 (-4, 3) B^1 (-1, 3) and C^1 (x, y) by a

Transformation M = $\begin{pmatrix} a & b \\ c & d \end{pmatrix}$

a) Find the: (i) Matrix M of the transformation

(ii) Coordinates of C_1

b) Triangle T^2 is the image of triangle T^1 under a reflection in the line y = x.

Find a single matrix that maps T and T_2

5. Triangles ABC is such that A is (2, 0), B (2, 4), C (4, 4) and A"B"C" is such that A" is (0, 2), B" (-4 – 10) and C "is (-4, -12) are drawn on the Cartesian plane

Triangle ABC is mapped onto A"B"C" by two successive transformations

R = $\begin{pmatrix} a & b \\ c & d \end{pmatrix}$ Followed by P = $\begin{pmatrix} 0 & -1 \\ -1 & 0 \end{pmatrix}$

(a) Find R

(b) Using the same scale and axes, draw triangles A'B'C', the image of triangle ABC under transformation R

Describe fully, the transformation represented by matrix R

6.   Triangle ABC is shown on the coordinates plane below

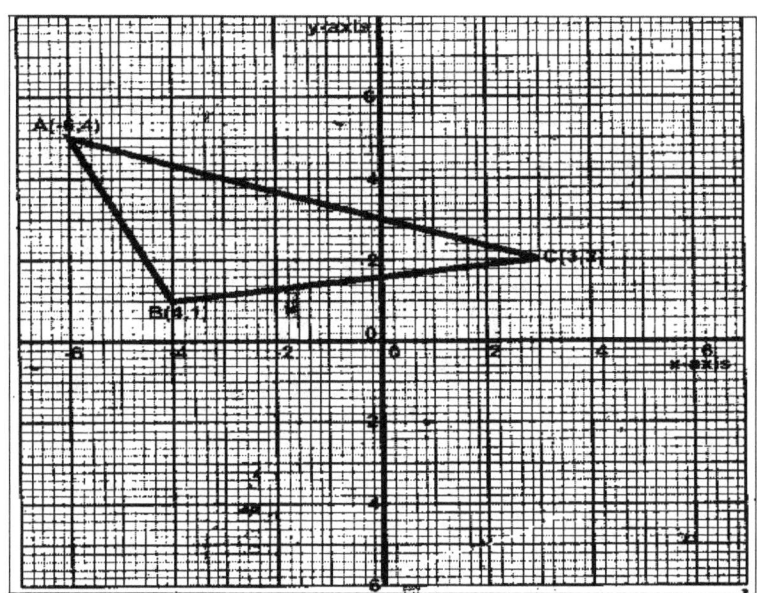

(a)   Given that A (-6, 5) is mapped onto A (6,-4) by a shear with y- axis invariant

        (i)   Draw triangle A'B'C', the image of triangle ABC under the shear

        (ii)   Determine the matrix representing this shear

(b)   Triangle A B C is mapped on to A" B" C" by a transformation defined by the

        matrix -1 $\begin{pmatrix} 0 & \\ 1\frac{1}{2} & -1 \end{pmatrix}$

        (i) Draw triangle A" B" C"

        (ii) Describe fully a single transformation that maps ABC onto A"B" C"

7.   Determine the inverse T⁻¹ of the matrix $\begin{pmatrix} 1 & 2 \\ 1 & -1 \end{pmatrix}$

       Hence find the coordinates to the point at which the two lines

       x + 2y = 7 and x - y =1

129

8. Given that $A = \begin{pmatrix} 0 & -1 \\ 3 & 2 \end{pmatrix}$ and $B = \begin{pmatrix} -1 & 0 \\ 2 & -4 \end{pmatrix}$

Find the value of x if

(i) $A - 2x = 2B$

(ii) $3x - 2A = 3B$

(iii) $2A - 3B = 2x$

9. The transformation R given by the matrix

$A = \begin{pmatrix} a & b \\ c & d \end{pmatrix}$ maps $\begin{pmatrix} 17 \\ 0 \end{pmatrix}$ to $\begin{pmatrix} 15 \\ 8 \end{pmatrix}$ and $\begin{pmatrix} 0 \\ 17 \end{pmatrix}$ to $\begin{pmatrix} -8 \\ 15 \end{pmatrix}$

(a)   Determine the matrix A giving a, b, c and d as fractions

(b)   Given that A represents a rotation through the origin determine the angle of rotation.

(c)   S is a rotation through 180 about the point (2, 3). Determine the image of (1, 0) under S followed by R.

CHAPTER THIRTY TWO: STATISTICS

1.   Every week the number of absentees in a school was recorded. This was done for 39 weeks these observations were tabulated as shown below

Number of absentees	0.3	4 -7	8 -11	12 - 15	16 - 19	20 – 23
(Number of weeks)	6	9	8	11	3	2

Estimate the median absentee rate per week in the school

2.   The table below shows high altitude wind speeds recorded at a weather station in a period of 100 days.

Wind speed (knots)	0 - 19	20 – 39	40 - 59	60-79	80- 99	100- 119	120-139	140-159	160-179
Frequency (days)	9	19	22	18	13	11	5	2	1

   (a)   On the grid provided draw a cumulative frequency graph for the data

   (b)   Use the graph to estimate

      (i)   The interquartile range

      (ii)   The number of days when the wind speed exceeded 125 knots

3.   Five pupils A, B, C, D and E obtained the marks 53, 41, 60, 80 and 56 respectively. The table below shows part of the work to find the standard deviation.

Pupil	Mark x	x - a	(x-a)^2
A	53	-5	
B	41	-17	
C	60	2	
D	80	22	
E	56	-2	

   (a)   Complete the table

   (b)   Find the standard deviation

4.	In an agricultural research centre, the length of a sample of 50 maize cobs were measured and recorded as shown in the frequency distribution table below.

Length in cm	Number of cobs
8 – 10	4
11 – 13	7
14 – 16	11
17 – 19	15
20 – 22	8
23 – 25	5

Calculate

    (a) The mean

    (b) (i) The variance

        (ii) The standard deviation

5.	The table below shows the frequency distribution of masses of 50 new- born calves in a ranch

Mass (kg)	Frequency
15 – 18	2
19- 22	3
23 – 26	10
27 – 30	14
31 – 34	13
35 – 38	6
39 – 42	2

(a)   On the grid provided draw a cumulative frequency graph for the data

(b)   Use the graph to estimate

     (i)   The median mass

     (ii)   The probability that a calf picked at random has a mass lying between 25 kg and 28 kg.

6. The table below shows the weight and price of three commodities in a given period

Commodity	Weight	Price Relatives
X	3	125
Y	4	164
Z	2	140

Calculate the retail index for the group of commodities.

7. The number of people who attended an agricultural show in one day was 510 men, 1080 women and some children. When the information was represented on a pie chart, the combined angle for the men and women was 216^0. Find the angle representing the children.

8. The mass of 40 babies in a certain clinic were recorded as follows:

Mass in Kg	No. of babies.
1.0 – 1.9	6
2.0 – 2.9	14
3.0 -3.9	10
4.0 – 4.9	7
5.0 – 5.9	2
6.0 – 6.9	1

Calculate

(a)   The inter – quartile range of the data.

(b)   The standard deviation of the data using 3.45 as the assumed mean.

9.   The data below shows the masses in grams of 50 potatoes

Mass (g)	25- 34	35-44	45 - 54	55- 64	65 - 74	75-84	85-94
No of potatoes	3	6	16	12	8	4	1

(a)   On the grid provide, draw a cumulative frequency curve for the data

(b)   Use the graph in (a) above to determine

   (i)   The 60th percentile mass

   (ii)   The percentage of potatoes whose masses lie in the range 53g to 68g

10.   The histogram below represents the distribution of marks obtained in a test.

The bar marked A has a height of 3.2 units and a width of 5 units. The bar marked B has

a height of 1.2 units and a width of 10 units

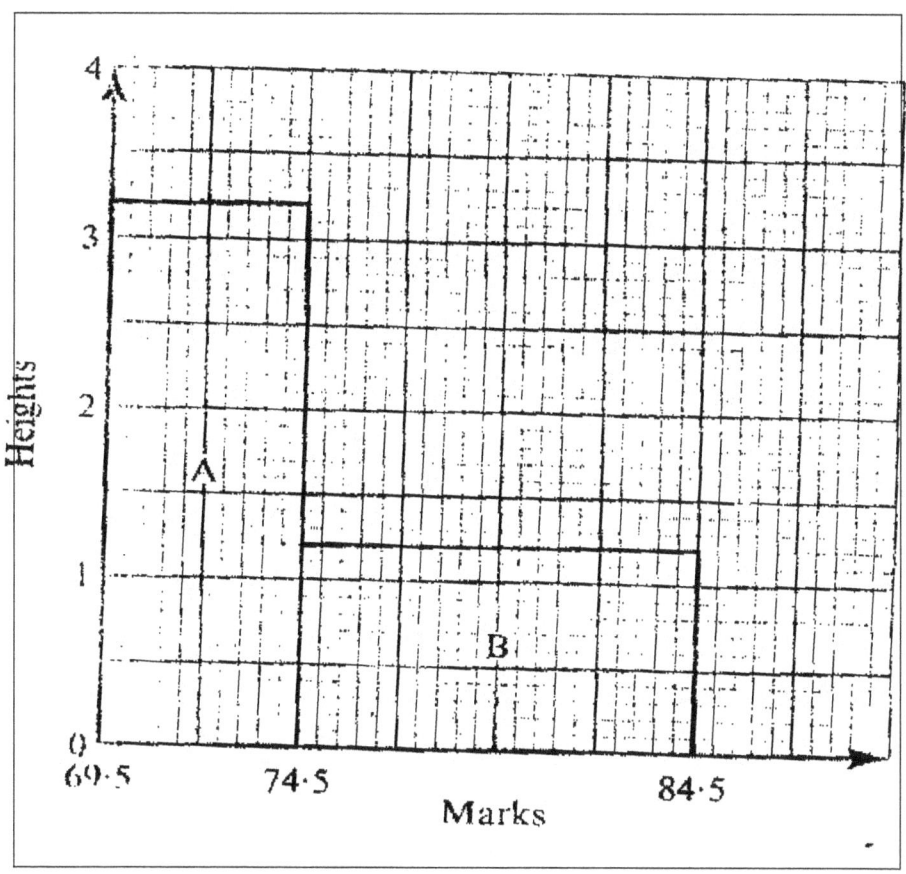

If the frequency of the class represented by bar B is 6, determine the frequency of the class represented by bar A.

11. A frequency distribution of marks obtained by 120 candidates is to be represented in a histogram. The table below shows the grouped marks. Frequencies for all the groups and also the area and height of the rectangle for the group 30 – 60 marks.

Marks	0-10	10-30	30-60	60-70	70-100
Frequency	12	40	36	8	24
Area of rectangle			180		
Height of rectangle			6		

(a) (i)   Complete the table

    (ii) On the grid provided below, draw the histogram

(b) (i)   State the group in which the median mark lies

    (ii) A vertical line drawn through the median mark divides the total area of the

        histogram into two equal parts

        Using this information or otherwise, estimate the median mark

12.   In an agriculture research centre, the lengths of a sample of 50 maize cobs were measured

and recorded as shown in the frequency distribution table below

Length in cm	Number of cobs
8 – 10	4
11- 13	7
14 – 16	11
17- 19	15
20 – 22	8
23- 25	5

Calculate

(a)   The mean

(b)   (i)   The variance

       (ii)   The standard deviation

12. The table below shows the frequency distribution of masses of 50 newborn calves in a ranch.

Mass (kg)	Frequency
15 – 18	2
19- 22	3
23 – 26	10
27 – 30	14
31- 34	13
35 – 38	6
39 – 42	2

(a)   On the grid provided draw a cumulative frequency graph for the data

(b)   Use the graph to estimate

    (i)   The median mass

    (ii)   The probability that a calf picked at random has a mass lying between 25 kg and 28 kg

14. The table shows the number of bags of sugar per week and their moving averages

Number of bags per week	340	330	X	343	350	345
Moving averages		331	332	y	346	

(a) Find the order of the moving average

(b) Find the value of X and Y axis

CHAPTER THIRTY THREE: LOC1

1. Using ruler and compasses only, construct a parallelogram ABCD such that AB = 10cm, BC = 7 cm and < ABC = 105⁰. Also construct the loci of P and Q within the parallel such that AP ≤ 4 cm, and BC ≤ 6 cm. Calculate the area within the parallelogram and outside the regions bounded by the loci.

2. Use ruler and compasses only in this question

    The diagram below shows three points A, B and D

    (a) Construct the angle bisector of acute angle BAD

    (b) A point P, on the same side of AB and D, moves in such a way that < APB = 22 ½ ⁰ construct the locus of P

    (c) The locus of P meets the angle bisector of < BAD at C measure < ABC

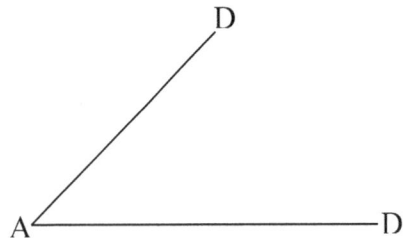

3. Use a ruler and a pair of compasses only for all constructions in this question.

    (a) On the line BC given below, construct triangle <ABC such that <ABC = 30⁰ and BA = 12 cm

    (b) Construct a perpendicular from A to meet BC produced at D. Measure CD

    (c) Construct triangle A'B'C' such that the area of triangle A'B'C is the three quarters of the area of triangle ABC and on the same side of BC as triangle ABC.

(d) Describe the lucus of A'

4. Use a ruler and compasses in this question. Draw a parallegram ABCD in which AB = 8 cm, BC = 6 cm and BAD = 75^0. By construction, determine the perpendicular distance between AB and CD.

5. In this question use a ruler and a pair of compasses.

   a) Line PQ drawn below is part of a triangle PQR. Construct the triangle PQR in which < QPR = 30^0 and line PR = 8 cm

   P|————————————————————————————|Q

   b) On the same diagram construct triangle PRS such that points S and Q are no the opposite sides of PR<PS = PS and QS = 8 cm

   C) A point T is on the a line passing through R and parallel to QS. If <QTS =90^0, locate possible positions of T and label them T_1 and T_2, Measure the length of T_1T_2.

6. (a) ABCD is a rectangle in which AB = 7.6 cm and AD = 5.2 cm. Draw the rectangle and construct the lucus of a point P within the rectangle such that P is equidistant from CB and CD   (3 marks)

   (b) Q is a variable point within the rectangle ABCD drawn in (a) above such that 60^0 $\leq$ < AQB$\leq 90^0$

   On the same diagram, construct and show the locus of point Q, by leaving unshaded, the region in which point Q lies.

7.	The figure below is drawn to scale. It represents a field in the shape of an equilateral triangle of side 80m

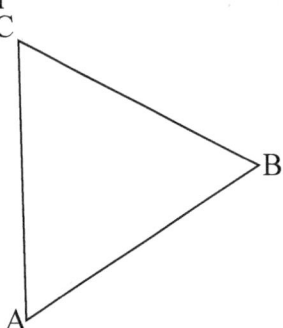

The owner wants to plant some flowers in the field. The flowers must be at most, 60m from A and nearer to B than to C. If no flower is to be more than 40m from BC, show by shading, the exact region where the flowers may be planted.

8.	In this question use a ruler and a pair of compasses only

In the figure below, AB and PQ are straight lines

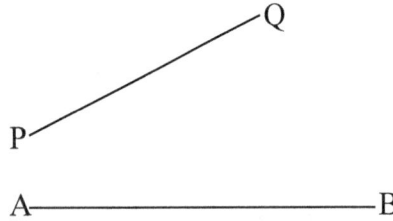

(a) Use the figure to:

(i)	Find a point R on AB such that R is equidistant from P and Q

(ii)	Complete a polygon PQRST with AB as its line of symmetry and hence measure the distance of R from TS.

(b) Shade the region within the polygon in which a variable point X must lie given that X satisfies the following conditions

1.	X is nearer to PT than to PQ

2.	RX is not more than 4.5 cm

3. $\angle$ PXT > 90^0

9.   Four points B, C, Q and D lie on same plane. Point B is 42 km due south – west of town

      Q. Point C is 50 km on a bearing of 560^0 from Q. Point D is equidistant from B, Q and C.

      (a)   Using the scale: 1 cm represents 10 km, construct a diagram showing the position

             of B, C, Q and D

      (b)   Determine the

             (i)   Distance between B and C

             (ii)   Bearing of D from B

10.   The diagram below represents a field PQR

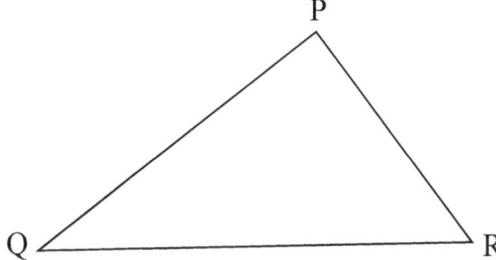

(a)   Draw the locus of point equidistant from sides PQ and PR

      (b)   Draw the locus of points equidistant from points P and R

      (c)   A coin is lost within a region which is near to point P than R and closer to side PR

             than to side PQ. Shade the region where the coin can be located.

12.   In the figure below, a line XY and three point A,B and C are as given. On the figure

      construct

      (a)   The perpendicular bisector if AB

141

(b)   A point P on the line XY such that angle APB = angle ACB

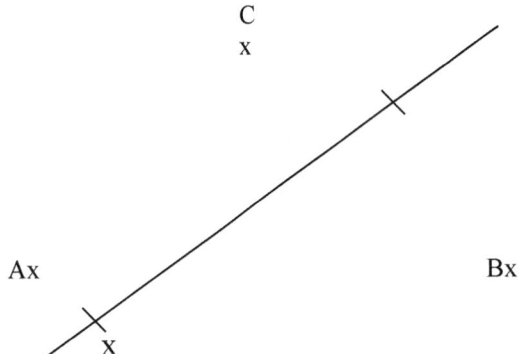

CHAPTER THIRTY FOUR: TRIGONOMETRY II

1. (a) Complete the table for the function $y = 2 \sin x$

X	0^0	10^0	20^0	30^0	40^0	50^0	60^0	70^0	80^0	90^0	100^0	110^0	120^0
Sin 3x	0	0.5000							-08660				
Y	0	1.00							-1.73				

   (b) (i) Using the values in the completed table, draw the graph of

   $y = 2 \sin 3x$ for $0^0 \leq x \leq 120^0$ on the grid provided

   (ii) Hence solve the equation $2 \sin 3x = -1.5$

2. Complete the table below by filling in the blank spaces

X^0	0^0	30^0	60^0	90^0	120^0	150^0	180^0	210^0	240^0	270^0	300^0	330^0	360^0
Cos x^0	1.00		0.50			-0.87		-0.87					
2 cos ½ x^0	2.00	1.93				0.52			-1.00				-2.00

Using the scale 1 cm to represent 30^0 on the horizontal axis and 4 cm to represent 1 unit on the vertical axis draw, on the grid provided, the graphs of $y = \cos x^0$ and $y = 2 \cos \frac{1}{2}$ x^0 on the same axis.

   (a) Find the period and the amplitude of $y = 2 \cos \frac{1}{2} x^0$

   (b) Describe the transformation that maps the graph of $y = \cos x^0$ on the graph of $y = 2 \cos \frac{1}{2} x^0$

2. (a) Complete the table below for the value of $y = 2 \sin x + \cos x$.

x	0^0	30^0	45^0	60^0	90^0	120^0	135^0	150^0	180^0	225^0	270^0	315^0	360^0
2 sin x	0		1.4	1.7	2	1.7	1.4	1	0		-2	-1.4	0
Cos x	1		0.7	0.5	0	-0.5	-0.7	-0.9	-1		0	0.7	1
y	1		2.1	2.2	2	1.2	0.7	0.1	-1		-2	-0.7	1

(b)   Using the grid provided draw the graph of y=2sin x + cos x for 0^0. Take 1cm

represent 30^0 on the x- axis and 2 cm to represent 1 unit on the axis.

(c)   Use the graph to find the range of x that satisfy the inequalities

2 sin x cos x > 0.5

4.   (a)   Complete the table below, giving your values correct to 2 decimal places.

X	0	10	20	30	40	50	60	70
Tan x	0							
2 x + 300	30	50	70	90	110	130	150	170
Sin (2x + 30^0)	0.50			1				

b)   On the grid provided, draw the graphs of y = tan x and y = sin (2x + 30^0) for $0^0 \leq$

x 70^0

Take scale: 2 cm for 100 on the x- axis

4 cm for unit on the y- axis

Use your graph to solve the equation tan x- sin (2x + 30^0) = 0.

5.   (a)   Complete the table below, giving your values correct to 2 decimal places

144

X^0	0	30	60	90	120	150	180
$2 \sin x^0$	0	1		2		1	
$1 - \cos x^0$			0.5	1			

(b) On the grid provided, using the same scale and axes, draw the graphs of

$y = \sin x^0$ and $y = 1 - \cos x^0 \leq x \leq 180^0$

Take the scale: 2 cm for 30^0 on the x- axis

2 cm for I unit on the y- axis

(c) Use the graph in (b) above to

(i) Solve equation

$2 \sin x^o + \cos x^0 = 1$

(ii) Determine the range of values x for which $2 \sin x^o > 1 - \cos x^0$

6. (a) Given that $y = 8 \sin 2x - 6 \cos x$, complete the table below for the missing values of y, correct to 1 decimal place.

X	0^0	15^0	30^0	45^0	60^0	75^0	90^0	105^0	120^0
$Y = 8 \sin 2x - 6 \cos x$	-6	-1.8		3.8	3.9	2.4	0		-3.9

(b) On the grid provided, below, draw the graph of $y = 8 \sin 2x - 6 \cos$ for

$0^0 \leq x \leq 120^0$

Take the scale 2 cm for 15^0 on the x- axis

2 cm for 2 units on the y – axis

(c) Use the graph to estimate

(i) The maximum value of y

(ii) The value of x for which $4 \sin 2x - 3 \cos x = 1$

7. Solve the equation $4 \sin (x + 30^0) = 2$ for $0 \leq x \leq 360^0$

8. Find all the positive angles not greater than 180^0 which satisfy the equation

$$\frac{\sin^2 x - 2 \tan x}{\cos x} = 0$$

9. Solve for values of x in the range $0^0 \leq x \leq 360^0$ if $3 \cos^2 x - 7 \cos x = 6$

10. Simplify $\frac{9 - y^2}{y}$ where $y = 3 \cos \theta$

11. Find all the values of Ø between 0^0 and 360^0 satisfying the equation $5 \sin \Theta = -4$

12. Given that $\sin (90 - x) = 0.8$. Where x is an acute angle, find without using mathematical tables the value of $\tan x^0$

13. Complete the table given below for the functions

$y = -3 \cos 2x^0$ and $y = 2 \sin (^{3x}/_2{}^0 + 30)$ for $0 \leq x \leq 180^0$

X^0	0^0	20^0	40^0	60^0	80^0	100^0	120^0	140^0	160^0	180^0
$-3\cos 2x^0$	-3.00	-2.30	-0.52	1.50	2.82	2.82	1.50	-0.52	-2.30	-3.00
$2 \sin (3 x^0 + 30^0)$	1.00	1.73	2.00	1.73	1.00	0.00	-1.00	-1.73	-2.00	-1.73

Using the graph paper draw the graphs of $y = -3 \cos 2x^0$ and $y = 2 \sin (^{3x}/_2{}^0 + 30^0)$

(a) On the same axis. Take 2 cm to represent 20^0 on the x- axis and 2 cm to represent one unit on the y – axis

(b) From your graphs. Find the roots of $3 \cos 2 x^0 + 2 \sin (^{3x}/_2{}^0 + 30^0) = 0$

14. Solve the values of x in the range $0^0 \le x \le 360^0$ if $3 \cos^2 x – 7\cos x = 6$

15. Complete the table below by filling in the blank spaces

x^0	0^0	30^0	60^0	90	1^0	150^0	180	210	240	270	300	330	360
$\cos x^0$	1.00		0.50			-0.87		-0.87					
$2\cos \frac{1}{2} x^0$	2.00	1.93					0.5						

Using the scale 1 cm to represent 30^0 on the horizontal axis and 4 cm to represent 1 unit on the vertical axis draw on the grid provided, the graphs of $y – \cos x^0$ and $y = 2 \cos \frac{1}{2}$ x^0 on the same axis

(a) Find the period and the amplitude of $y = 2 \cos \frac{1}{2} x^0$

Ans. Period $= 720^0$. Amplitude $= 2$

(b) Describe the transformation that maps the graph of $y = \cos x^0$ on the graph of $y = 2 \cos \frac{1}{2} x^0$

CHAPTER THIRTY FIVE: THREE DIMENSIONAL

GEOMETRY

1. The diagram below shows a right pyramid VABCD with V as the vertex. The base of the pyramid is rectangle ABCD, WITH ab = 4 cm and BC= 3 cm. The height of the pyramid is 6 cm.

2

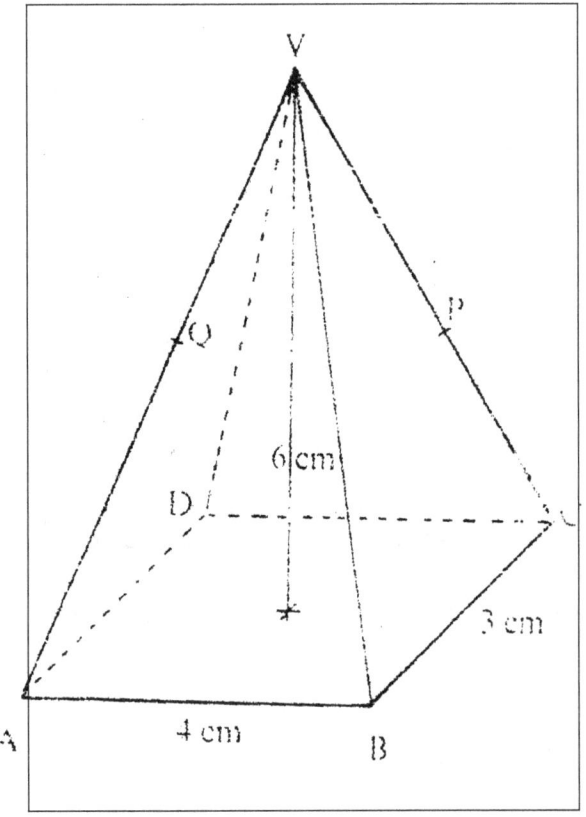

(a)   Calculate the

    (i)   Length of the projection of VA on the base

    (ii)   Angle between the face VAB and the base

(b)   P is the mid- point of VC and Q is the mid – point of VD.

    Find the angle between the planes VAB and the plane ABPQ

2.   The figure below represents a square based solid with a path marked on it.

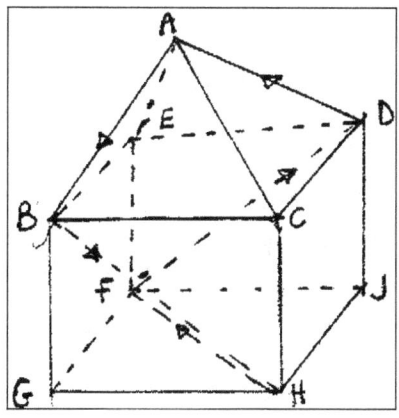

Sketch and label the net of the solid.

3.   The diagram below represents a cuboid ABCDEFGH in which FG= 4.5 cm, GH = 8 cm

and HC = 6 cm

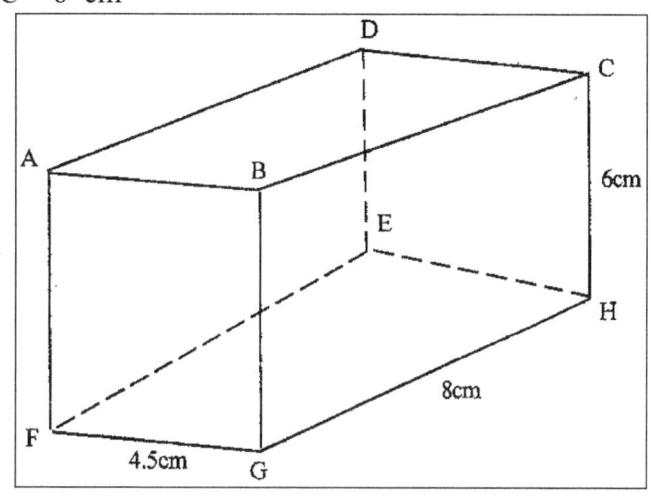

Calculate:

(a) The length of FC

(b) (i) The size of the angle between the lines FC and FH

      (ii) The size of the angle between the lines AB and FH

(c) The size of the angle between the planes ABHE and the plane FGHE

4.   The base of a right pyramid is a square ABCD of side 2a cm. The slant edges VA, VB, VC and VD are each of length 3a cm.

(a)   Sketch and label the pyramid

(b)   Find the angle between a slanting edge and the base

5.   The triangular prism shown below has the sides AB = DC = EF = 12 cm. the ends are equilateral triangles of sides 10cm. The point N is the mid point of FC.

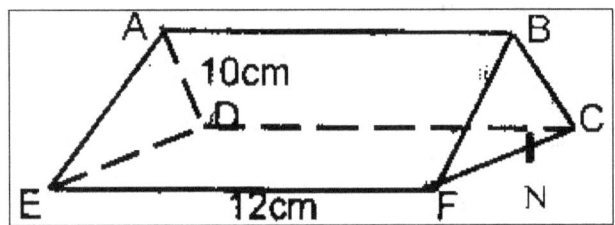

Find the length of:

(a)   (i)   BN

       (ii)   EN

(b)   Find the angle between the line EB and the plane CDEF

CHAPTER THIRTY SIX: LATITUDES AND LONGITUDES

1. An aeroplane flies from point A (1^0 15'S, 37^0 E) to a point B directly North of A. the arc

    AB subtends an angle of 45^0 at the center of the earth. From B, aeroplanes flies due west

    two a point C on longitude 23^0 W.)

    (Take the value of π $^{22}/_7$ as and radius of the earth as 6370km)

    (a)   (i)   Find the latitude of B

           (ii)   Find the distance traveled by the aeroplane between B and C

    (b)   The aeroplane left at 1.00 a.m local time. When the aeroplane was leaving B, hat

           was the local time at C?

2. The position of two towns X and Y are given to the nearest degree as X (45^0 N, 10^0W)

    and Y (45^0 N, 70^0W)

    Find

    (a)   The distance between the two towns in

           (i)   Kilometers (take the radius of the earth as 6371)

           (ii)   Nautical miles (take 1 nautical mile to be 1.85 km)

    (b)   The local time at X when the local time at Y is 2.00 pm.

3. A plane leaves an airport A (38.5^0N, 37.05^0W) and flies dues North to a point B on

    latitude 52^0N.

    (a)   Find the distance covered by the plane

    (b)   The plane then flies due east to a point C, 2400 km from B. Determine the

           position of C

           Take the value π of as $^{22}/_7$ and radius of the earth as 6370 km

4. A plane flying at 200 knots left an airport A (30^0S, 31^0E) and flew due North to an airport B (30^0N, 31^0E)

    (a) Calculate the distance covered by the plane, in nautical miles

    (b) After a 15 minutes stop over at B, the plane flew west to an airport C (30^0N, 13^0E) at the same speed.

    Calculate the total time to complete the journey from airport C, though airport B.

5. Two towns A and B lie on the same latitude in the northern hemisphere.

    When its 8 am at A, the time at B is 11.00 am.

    a) Given that the longitude of A is 15^0 E find the longitude of B.

    b) A plane leaves A for B and takes $3^1/_2$ hours to arrive at B traveling along a parallel of latitude at 850 km/h. Find:

        (i) The radius of the circle of latitude on which towns A and B lie.

        (ii) The latitude of the two towns (take radius of the earth to be 6371 km)

6. Two places A and B are on the same circle of latitude north of the equator. The longitude of A is 118^0W and the longitude of B is 133^0 E. The shorter distance between A and B measured along the circle of latitude is 5422 nautical miles.

    Find, to the nearest degree, the latitude on which A and B lie

7. (a) A plane flies by the short estimate route from P (10^0S, 60^0 W) to Q (70^0 N, 120^0 E) Find the distance flown in km and the time taken if the aver age speed is 800 km/h.

    (b) Calculate the distance in km between two towns on latitude 50^0S with long longitudes and 20^0 W. (take the radius of the earth to be 6370 km)

8. Calculate the distance between M (30^0N, 36^0E) and N (30^0 N, 144^0 W) in nautical miles.

   (i)   Over the North Pole

   (ii) Along the parallel of latitude 30^0 N

9. (a) A ship sailed due south along a meridian from 12^0 N to $10^0 30$'S. Taking the earth to be a sphere with a circumference of 4×10^4 km, calculate in km the distance traveled by the ship.

   (b) If a ship sails due west from San Francisco (37^0 47'N, 122^0 26'W) for distance of 1320 km. Calculate the longitude of its new position (take the radius of the earth to be 6370 km and $\pi = 22/7$).

CHAPTER THIRTY SEVEN: LINEAR PROGRAMMING

1. A school has to take 384 people for a tour. There are two types of buses available, type X and type Y. Type X can carry 64 passengers and type Y can carry 48 passengers. They have to use at least 7 buses.

    (a) Form all the linear inequalities which will represent the above information.

    (b) On the grid [provide, draw the inequalities and shade the unwanted region.

    (c) The charges for hiring the buses are

    Type X: Kshs 25,000

    Type Y Kshs 20,000

    Use your graph to determine the number of buses of each type that should be hired to minimize the cost.

2. An institute offers two types of courses technical and business courses. The institute has a capacity of 500 students. There must be more business students than technical students but at least 200 students must take technical courses. Let x represent the number of technical students and y the number of business students.

    (a) Write down three inequalities that describe the given conditions

    (b) On the grid provided, draw the three inequalities

    (c) If the institute makes a profit of Kshs 2, 500 to train one technical students and Kshs 1,000 to train one business student, determine

        (i) The number of students that must be enrolled in each course to maximize the profit

        (ii) The maximum profit.

3. A draper is required to supply two types of shirts A and type B.

The total number of shirts must not be more than 400. He has to supply more type A than of type B however the number of types A shirts must be more than 300 and the number of type B shirts not be less than 80.

Let x be the number of type A shirts and y be the number of types B shirts.

(a) Write down in terms of x and y all the linear inequalities representing the information above.

(b) On the grid provided, draw the inequalities and shade the unwanted regions

(c) The profits were as follows

Type A: Kshs 600 per shirt

Type B: Kshs 400 per shirt

    (i)   Use the graph to determine the number of shirts of each type that should be made to maximize the profit.

    (ii)   Calculate the maximum possible profit.

4.   A diet expert makes up a food production for sale by mixing two ingredients N and S. One kilogram of N contains 25 units of protein and 30 units of vitamins. One kilogram of S contains 50 units of protein and 45 units of vitamins. The foiod is sold in small bags each containing at least 175 units of protein and at least 180 units of vitamins. The mass of the food product in each bag must not exceed 6kg.

If one bag of the mixture contains x kg of N and y kg of S

(a) Write down all the inequalities, in terms of x and representing the information above

(b) On the grid provided draw the inequalities by shading the unwanted regions

(c) If one kilogram of N costs Kshs 20 and one kilogram of S costs Kshs 50, use the graph to determine the lowest cost of one bag of the mixture.

5. Mwanjoki flying company operates a flying service. It has two types of aeroplanes. The smaller one uses 180 litres of fuel per hour while the bigger one uses 300 litres per hour. The fuel available per week is 18,000 litres. The company is allowed 80 flying hours per week.

(a) Write down all the inequalities representing the above information

(b) On the grid provided on page 21, draw all the inequalities in (a) above by shading the unwanted regions

(c) The profits on the smaller aeroplane is Kshs 4000 per hour while that on the bigger one is Kshs. 6000 per hour. Use your graph to determine the maximum profit that the company made per week.

6. A company is considering installing two types of machines. A and B. The information about each type of machine is given in the table below.

Machine	Number of operators	Floor space	Daily profit
A	2	$5m^2$	Kshs 1,500
B	5	$8m^2$	Kshs 2,500

The company decided to install x machines of types A and y machines of type B

(a) Write down the inequalities that express the following conditions

i. The number of operators available is 40

ii. The floor space available is $80m^2$

iii. The company is to install not less than 3 type of A machine

iv. The number of type B machines must be more than one third the number of type A machines

(b) On the grid provided, draw the inequalities in part (a) above and shade the unwanted region.

(c) Draw a search line and use it to determine the number of machines of each type that should be installed to maximize the daily profit.

CHAPTER THIRTY EIGHT: CALCULUS

1.   The shaded region below represents a forest. The region has been drawn to scale where 1 cm represents 5 km. Use the mid – ordinate rule with six strips to estimate the area of forest in hectares.

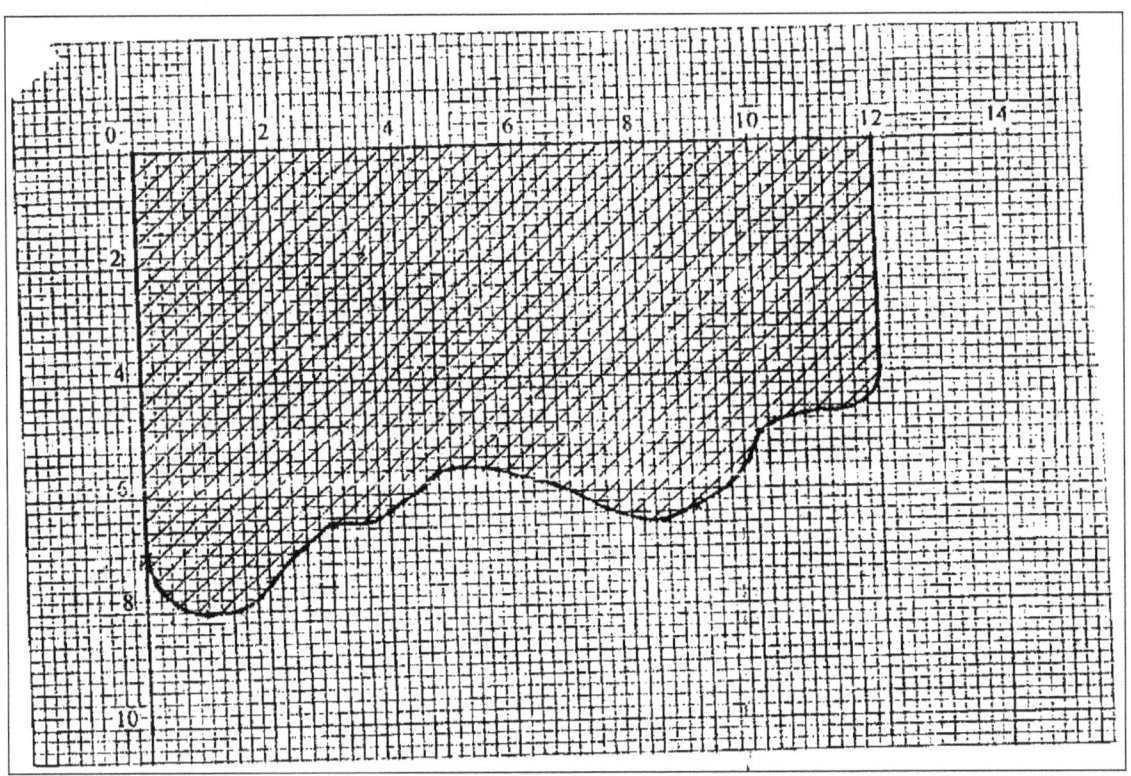

2.   Find the area bounded by the curve $y=2x^3 - 5$, the x-axis and the lines x=2 and x=4.

3.   Complete the table below for the function $y=3x^2 - 8x + 10$

x	0	2	4	6	8	10
y	10	6		70		230

Using the values in the table and the trapezoidal rule, estimate the area bounded by the curve $y= 3x^2 - 8x + 10$ and the lines y=0, x=0 and x=10.

4. Use the trapezoidal rule with intervals of 1 cm to estimate the area of the shaded region below

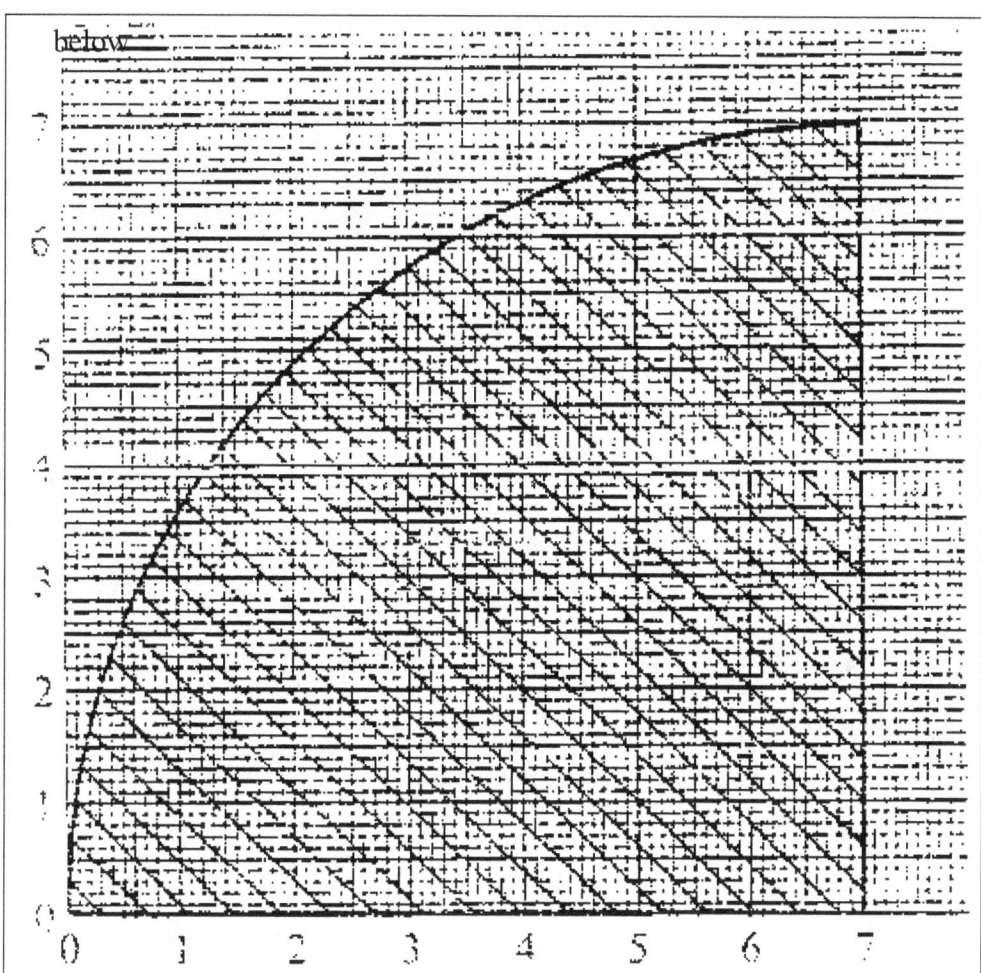

5. (a) Find the value of x at which the curve $y = x - 2x^2 - 3$ crosses the x- axis

   (b) Find $\int(x^2 - 2x - 3)\,dx$

   (c) Find the area bounded by the curve $y = x^2 - 2x - 3$, the axis and the lines $x = 2$ and

   $x = 4$.

6. The graph below consists of a non- quadratic part $(0 \leq x \leq 2)$ and a quadrant part $(2 \leq x$

8). The quadratic part is $y = x^2 - 3x + 5, 2 \leq x \leq 8$

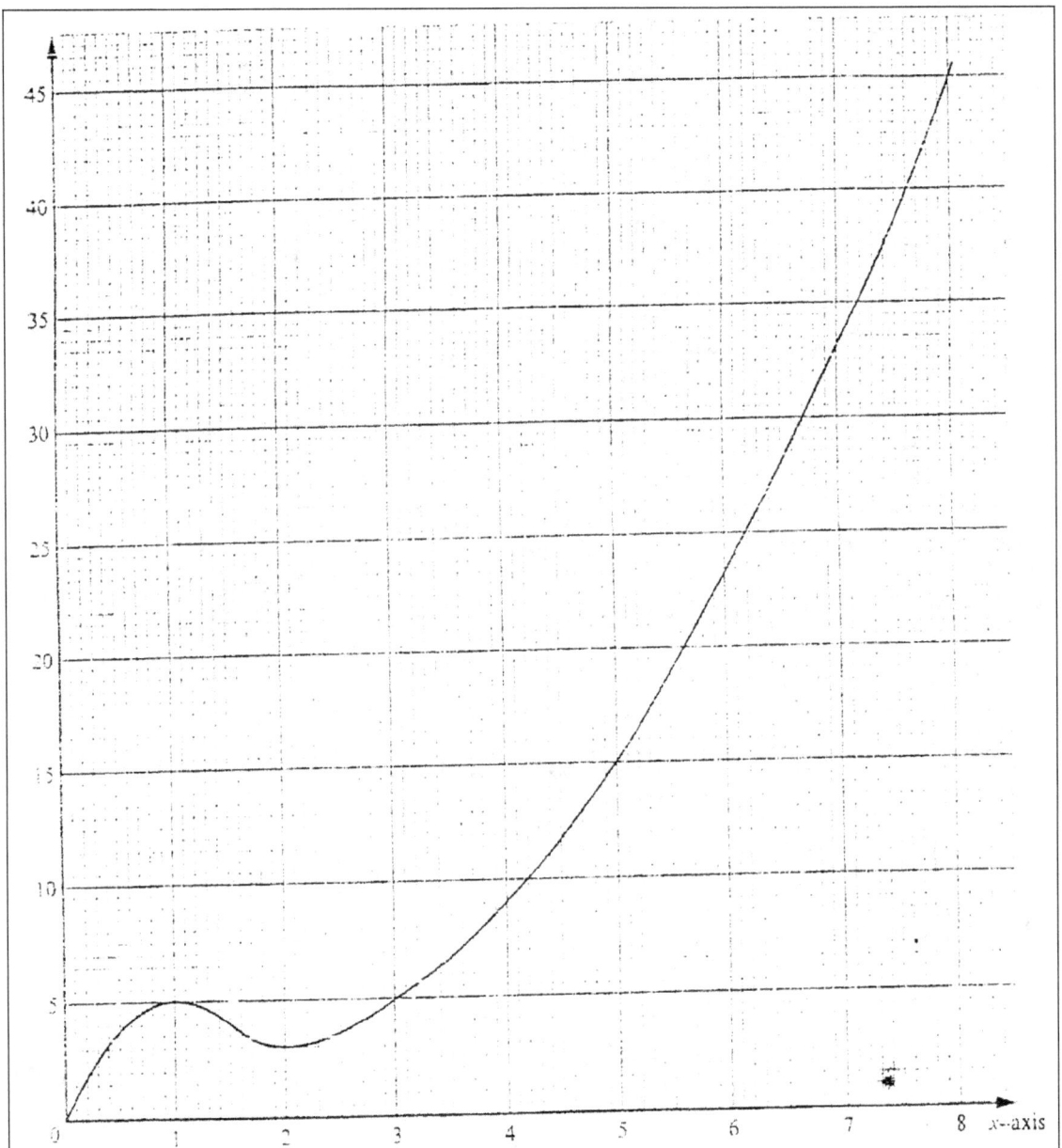

(a) Complete the table below

x	2	3	4	5	6	7	8
y	3						

(b) Use the trapezoidal rule with six strips to estimate the area enclosed by the curve, x = axis and the line x = 2 and x = 8

(c) Find the exact area of the region given in (b)

(d) If the trapezoidal rule is used to estimate the area under the curve between x = 0 and x = 2, state whether it would give an under- estimate or an over-estimate. Give a reason for your answer.

7. Find the equation of the gradient to the curve $Y = (x^{-2} + 1)(x - 2)$ when x = 2

8. The distance from a fixed point of a particular in motion at any time t seconds is given by

$$S = \frac{t^3 - 5t^2 + 2t + 5}{2t^2}$$

Find its:

(a) Acceleration after 1 second

(b) Velocity when acceleration is Zero

9. The curve of the equation $y = 2x + 3x^2$, has x = -2/3 and x = 0 and x intercepts.

The area bounded by the axis x = -2/3 and x = 2 is shown by the sketch below.

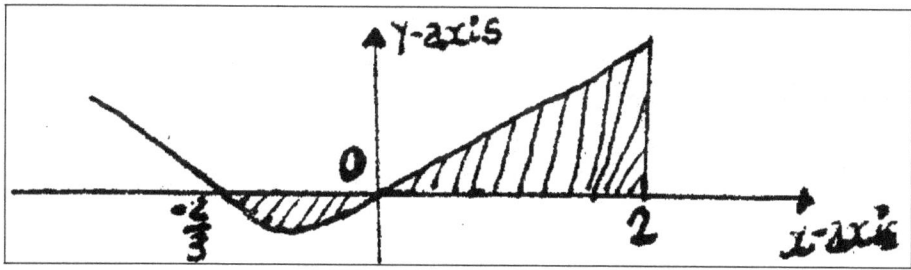

Find:

(a) $(2x + 3x^2)\,dx$

(b) The area bounded by the curve x – axis, $x = -\frac{2}{3}$ and $x = 2$

10.   A particle is projected from the origin. Its speed was recorded as shown in the table below

Time (sec)	0	5	10	15	20	25	39	35
Speed (m/s)	0	2.1	5.3	5.1	6.8	6.7	4.7	2.6

Use the trapezoidal rule to estimate the distance covered by the particle within the 35 seconds.

11.   (a)   The gradient function of a curve is given by $\dfrac{dy}{dx} = 2x^2 - 5$

Find the equation of the curve, given that $y = 3$, when $x = 2$

(b)   The velocity, vm/s of a moving particle after seconds is given:

$v = 2t^3 + t^2 - 1$. Find the distance covered by the particle in the interval $1 \le t \le 3$

12.   Given the curve $y = 2x^3 + \frac{1}{2}x^2 - 4x + 1$. Find the:

i)   Gradient of curve at $\{1, -\frac{1}{2}\}$

ii)   Equation of the tangent to the curve at $\{1, -\frac{1}{2}\}$

13. The diagram below shows a straight line intersecting the curve $y = (x-1)^2 + 4$

At the points P and Q. The line also cuts x-axis at (7, 0) and y axis at (0, 7)

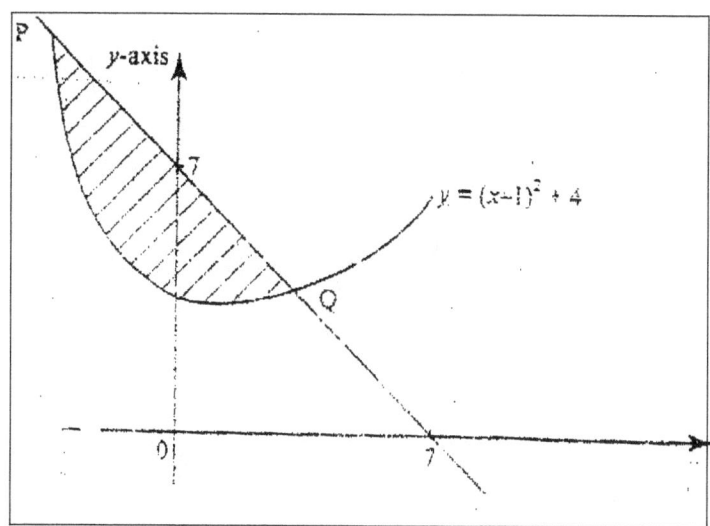

a) Find the equation of the straight line in the form y = mx +c.

b) Find the coordinates of p and Q.

c) Calculate the area of the shaded region.

14. The acceleration, a ms⁻², of a particle is given by $a = 25 - 9t^2$, where t in seconds after the particle passes fixed point O.

If the particle passes O, with velocity of 4 ms⁻¹, find

(a) An expression of velocity V, in terms of t

(b) The velocity of the particle when t = 2 seconds

15. A curve is represented by the function $y = \frac{1}{3} x^3 + x^2 - 3x + 2$

(a) Find: $\dfrac{dy}{dx}$

(b) Determine the values of y at the turning points of the curve

$$y = \frac{1}{3}x^3 + x^2 - 3x + 2$$

163

(c)   In the space provided below, sketch the curve of $y = \frac{1}{3} x^3 + x^2 - 3x + 2$

16.   A circle centre O, ha the equation $x^2 + y^2 = 4$. The area of the circle in the first quadrant is divided into 5 vertical strips of width 0.4 cm

(a)   Use the equation of the circle to complete the table below for values of y correct to 2 decimal places

X	0	0.4	0.8	1.2	1.6	2.0
Y	2.00			1.60		0

(b)   Use the trapezium rule to estimate the area of the circle

17.   A particle moves along straight line such that its displacement S metres from a given point is $S = t^3 - 5t^2 + 4$ where t is time in seconds

Find

(a) The displacement of particle at $t = 5$

(b) The velocity of the particle when $t = 5$

(c) The values of t when the particle is momentarily at rest

(d) The acceleration of the particle when $t = 2$

18.   The diagram below shows a sketch of the line $y = 3x$ and the curve $y = 4 - x^2$ intersecting at points P and Q.

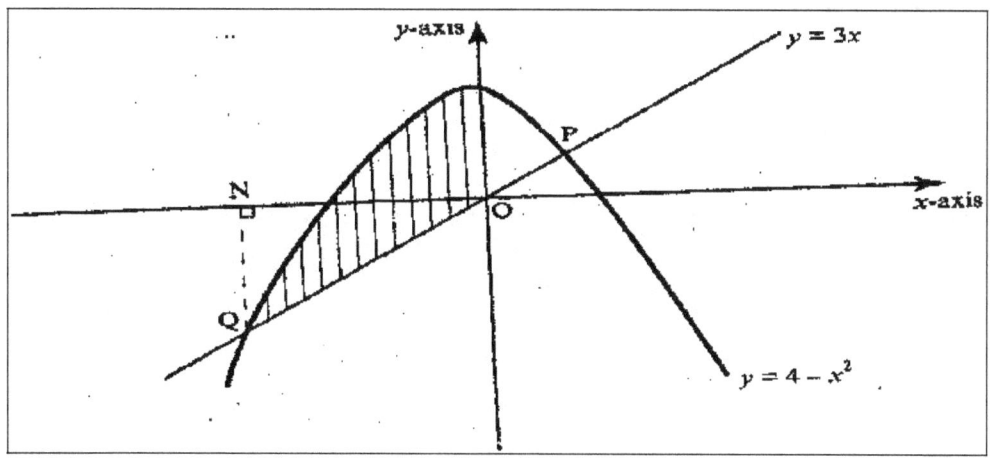

(a)   Find the coordinates of P and Q

(b)   Given that QN is perpendicular to the x- axis at N, calculate

    (i)   The area bounded by the curve $y = 4 - x^2$, the x- axis and the line QN

    (ii)   The area of the shaded region that lies below the x- axis

    (iii)   The area of the region enclosed by the curve $y = 4-x^2$, the line

            $y - 3x$ and the y-axis.

19.   The gradient of the tangent to the curve $y = ax^3 + bx$ at the point $(1, 1)$ is -5

    Calculate the values of a and b.

20.   The diagram on the grid below represents as extract of a survey map showing

    two adjacent plots belonging to Kazungu and Ndoe.

    The two dispute the common boundary with each claiming boundary along different

    smooth curves coordinates (x, y) and (x, y_2) in the table below, represents points on the

    boundaries as claimed by Kazungu Ndoe respectively.

165

x	0	1	2	3	4	5	6	7	8	9
y_1	0	4	5.7	6.9	8	9	9.8	10.6	11.3	12
y_2	0	0.2	0.6	1.3	2.4	3.7	5.3	7.3	9.5	12

(a) On the grid provided above draw and label the boundaries as claimed by Kazungu and Ndoe.

(b) (i) Use the trapezium rule with 9 strips to estimate the area of the section of the land in dispute

   (ii) Express the area found in b (i) above, in hectares, given that 1 unit on each axis represents 20 metres

21. The gradient function of a curve is given by the expression $2x + 1$. If the curve passes through the point (-4, 6);

(a) Find:

   (i) The equation of the curve

   (ii) The vales of x, at which the curve cuts the x- axis

(b) Determine the area enclosed by the curve and the x- axis

22. A particle moves in a straight line through a point P. Its velocity v m/s is given by $v = 2 - t$, where t is time in seconds, after passing P. The distance s of the particle from P when t = 2 is 5 metres. Find the expression for s in terms of t.

23. Find the area bonded by the curve $y = 2x - 5$ the x-axis and the lines x=2 and x = 4.

23. Complete the table below for the function

166

$Y = 3x^2 - 8x + 10$

X	0	2	4	6	8	10
Y	10	6	-	70	-	230

Using the values in the table and the trapezoidal rule, estimate the area bounded by the curve $y = 3x^2 - 8x + 10$ and the lines $y - 0$, $x = 0$ and $x = 10$

24.　(a)　Find the values of x which the curve $y = x^2 - 2x - 3$ crosses the axis

　　　(b)　Find $(x^2 - 2x - 3) \, dx$

　　　(c)　Find the area bounded by the curve $Y = x^2 - 2x - 3$. The x – axis and the lines $x = 2$ and $x = 4$

25.　Find the equation of the tangent to the curve $y = (x + 1)(x - 2)$ when $x = 2$

26.　The distance from a fixed point of a particle in motion at any time t seconds is given by s

$= t - \frac{5}{2}t^2 + 2t + s$ metres

Find its

　　(a)　Acceleration after t seconds

　　(b)　Velocity when acceleration is zero

27.　The curve of the equation $y = 2x + 3x^2$, has $x = -\frac{2}{3}$ and $x = 0$, as x intercepts. The area bounded by the curve, x – axis, $x = -\frac{2}{3}$ and $x = 2$ is shown by the sketch below.

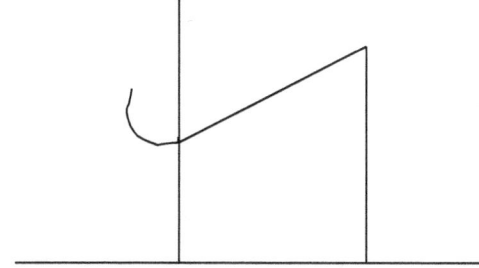

(a)   Find $\int(2x + 3x^2) \, dx$

(b)   The area bounded by the curve, x axis $x = -\frac{2}{3}$ and $x = 2$

28.   A curve is given by the equation $y = 5x^3 - 7x^2 + 3x + 2$

Find the

(a)   Gradient of the curve at $x = 1$

(b)   Equation of the tangent to the curve at the point $(1, 3)$

29.   The displacement x metres of a particle after t seconds is given by $x = t^2 - 2t + 6$, $t > 0$

(a)   Calculate the velocity of the particle in m/s when $t = 2s$

(b)   When the velocity of the particle is zero,

Calculate its

(i)   Displacement

(ii)   Acceleration

30.   The displacement s metres of a particle moving along a straight line after t seconds is given by $s = 3t + \frac{3}{2}t^2 - 2t^3$

(a)   Find its initial acceleration

(b)   Calculate

(i)   The time when the particle was momentarily at rest.

(ii)   Its displacement by the time it comes to rest momentarily when

$t = 1$ second, $s = 1\frac{1}{2}$ metres when $t = \frac{1}{2}$ seconds

168

(c)   Calculate the maximum speed attained

SOLUTIONS

CHAPTER ONE: NUMBERS

1.   1000 | 0.0064
            \ 100

   1000   (0.08)
              10

   1000 x 0.008

   =8

2.   (a) $\dfrac{-8 \div 2 + 12 \times 9 - 4 \times 6}{56 \div 7 \times 2}$

   $\dfrac{-4 + 108 - 24}{16}$

   = 80/16

   = 5

3. $\dfrac{46 - 3}{-2} = 23 - 1$

       3 = 24

4.   Mliwa: $^3/_8 \times {}^2/_3 x = {}^1/_4 x$

   Amina: $x - (^1/_3 + {}^1/_4)x = {}^5/_{12} x$

               $^5/_{12} x - {}^1/_4 x = 40{,}000$

               $^2/_{12} x = 40{,}000$

$$X = 240,000$$

5. $\quad \dfrac{+4 \times 4 - (-20)}{-6(6 \div 3) + (-6)} = \dfrac{4 \times 4 + 20}{-6 \times 2 - 6} = \dfrac{36}{-18}$

$\quad = 2$

6. $\quad \dfrac{384.\ 16 \times 0.625}{96.04}$

$$\sqrt{\dfrac{2^4 \times 7^4 \times 10^{-2} \times 5^4 \times 10^{-4}}{2^2 \times 7^4 \times 10^{-2}}}$$

$$\sqrt{2^2 \times 5^4 \times 10^{-4}}$$

$\qquad = 2 \times 5^2 \times 10^{-2}$

$\qquad\quad = 0.5$

7 $\quad 1/x + 1/x + 5 = \,^1/_6$

$\quad 6(x + 5) + 6x = x(x + 5)$

$\quad X2 - 7x - 30 = 0$

$\quad (x - 10)(x + 3) = 0$

$\quad X = 10, -3$

$\quad$ Onduso takes 10 days

8. $(1470)^2 = [2 \times 3 \times 5 \times 7^2]^8$

$\quad \sqrt{7056} \qquad \sqrt{(2^4 \times 3^2 \times 7^2)}$

$$= 2^2 \times 3^2 \times 5^2 \times 7^4$$

$$2^2 \times 3 \times 7$$

$$= 3 \times 5^2 \times 7^3$$

9. $\frac{3}{4} + 1\,\frac{5}{7} \div \frac{4}{7}$ of $2\,\frac{1}{3}$

$$(^{13}/_7 - \,^5/_8) \times \,^2/_3$$

$$\underline{\frac{3}{4} + 9/7}$$

$$^{45}/_{56} \times \,^2/_3$$

$$^{57}/_{28} \times \,^{28}/_{15} \text{ or } ^{399}/_{196} \times \,^{28}/_{15}$$

10. A and B opened for 1 hr

$$^1/_3 + \,^1/_6 = \frac{1}{2}$$

A,B,C opened for 1 hr

$$\frac{1}{2} - \,^1/_8 = \,^3/_8$$

Time taken to fill the tank when all pieces are opened $= \frac{1}{2} \times \,^2/_3 + 1$

21/3 hr

11. $^4/_9 (45 + W) = 10 + W$

$$4 (45 + w) = 9 (10 + w)$$

$$180 + 4W = 90 + 8 w$$

$$5w = 90$$

$$W = 18$$

12. $\sqrt{91125}$

$\sqrt{2025}$

$^{45}/_{45} = 1$

13. (a)   7532

    (b)   500

14. $\dfrac{0.0084 \times 1.23 \times 3.5}{2.87 \times 0.056}$

    $\dfrac{84 \times 123 \times 35}{287 \times 56 \times 100}$

    $= {}^{9}/_{40}$

15. $^{14}/_{7}$

16. 3

17. $^{4}/_{5}$ or 0.8

18. $^{1}/_{27}$

19. 11. 25

20. 30

21. $6\,{}^{5}/_{18}$

22. -17

23. $1\frac{5}{11}$

24.   a = 38, b = 225

25.   GCD = xy^2, xy^2 (x- 2y) x + 2y)

26. $\frac{9}{4}$

27.   48

CHAPTER TWO: ALGEBRAIC EXPRESSIONS

1. Let Ali be a goats

   $A + a + 2 + 3 (a + 2) + a + 2 + 3 (a + 2) – 10$

   $9a + b$

   $9a + 6 = 17 \times 3$

   $9a = 45$

   $a = 5$

   Odupoy sold $28 – 10 = 18$ goats

2. $yx + 3 yz = 2x - z$

   $X (2-y) = 3yz + z)$

   $X = \dfrac{z (3y + 1)}{2 - y}$

3. $3x^2 – 3xy + xy – y^2$

   $3x (x-y) + y (x – y)$

   $(x-y) (3x + y)$

4. $\dfrac{3 (x-1) –(2x + 1)}{3x} = \dfrac{3x – 3 – 2 x -1}{3x}$

   $\dfrac{X – 4}{3x}$

   $\dfrac{X -4}{3x} = \dfrac{2}{3}$

   $3x – 12 = 6x$

X = -4

5. $(a + b)(a - b)$

$(2557 - 2547)(2557 + 2547)$

510×10

51040

6. $\underline{(p + q)(p + q)}$

$P^2(p + q) - q^2(p + q)$

$\underline{(p + q)(p + q)}$ $\underline{1}$

$(p + q)(p - q)(p + q)$ =   p-q

7. $yx + 3yz = 2x - z$

$Yx - 2x = -3yz - z$

$X(y-2) = 3yz - z$

$X = -\underline{3x - z}$

         Y -2

8. $^1/_4x = {}^5/_6x - 7$

     $\underline{\quad a \quad}$ + $\underline{\quad b \quad}$

9. $2(a + b)$ $2(a - b)$

10. $(7x - 1)(4x + 1)$

11.   Ali's age = 16 yrs. Juma's age = 42 yrs

12.   Trouser 150, shirt cost30

CHAPTER THREE: RATES, RATIO PERCENTAGES AND PROPORTION

1. $\underline{(4 \times 21) + (3 \times 42)} = 30$

    7

    $\underline{130} \times 30 = 39$

    100

2. $\underline{27 \times 4 \times 60} = 3.6$

    60×30

    Height = 23.6 cm

3. (a) (i) Total collected Kshs 80 x 25 x 6

    Kshs 12000

    (ii) Net profit = 12000 – (1500 + 200 + 150 + 4000

    Kshs 12000 – 5850 = Kshs 6250

    (b) The days collection = Kshs $\underline{80}$ x 12000

    100

    = Kshs 9600

    Net profit = Kshs 9600- 5850

    Kshs 3750

    Shares = $^{25}/_5$ x 3750 or $^3/_5$ x 3750

    Kshs 1500 and Kshs. 2250

4. $^{3.5}/_{100}$ x 50 = 1.75

    (a) 4.75 x 30 = 1.425

Total = 3.175 kg

(ii) $\underline{3.175} \times 100 = 3.9688$
     80

$= 3.969\%$

(b)   No. of fat kg $= \,^{x}/_{50} \times 100 = 4$

          $X = 2$ kg fat

Milk

Kg of A = y

Kg of B = 50 – y

$\underline{3/5y} + \underline{4.75\ (50 - y)} = 2$
100   100

$3.5y + 237.5 - 4.75 = 200$

$1.2y = 37.5$

$Y = \underline{37.7}$
    1.25

$Y = 30$

A= 30 kg

B = 20 kg

B≥ 20 kg

5.   (a)   240 x 12000

= Kshs 2,880, 000

(b) (i) New price = $^{125}/_{100}$ x 12000

= Kshs. 15,000

New no of sets = $^{90}/_{100}$ x 240 = 216

Amount from sale = 216 x 15,000

= Kshs 3, 240,000

Increase = 3, 240, 000 – 2, 880,000

= 360, 000

% increase = $\underline{360, 000 \text{ x } 100}$ = 12.5%

2,880, 000

(ii) $^{16}/_{15}$ x 15,000   = Kshs 16,000

(c) Let the no of sets sold in 2003 be x

16000 x = 2,880,000

X = $\underline{2, 880, 000}$

16,000

$P\% = \dfrac{240 - 180}{240} \times 100 = 25\%$

$\therefore p = 25$

6.   (a)   Initial volume of alcohol

$= {}^{60}/_{100} \times 80$

New volume of solution $= (80 + x)$ ltrs)

$\dfrac{48}{(80 + x)} = \dfrac{40}{100}$

$4800 = 3200 + 40x$

$40x = 1600$

$X = 40$ ltrs

(b) New volume of solution

$80 + 40 + 30 = 150$ ltrs

$48/150 \times 100 = 32$

% age of alcohol $= 32\%$

(b)   in 5 lts

32% of 5 $= 1.6$ ltrs of alcohol

68% of 5 $= 3.4$ ltrs of water

In 2 ltrs   60% of 2 $= 1.2$ lts of alcohol

40% of 2 = 0.8 ltrs of water

In final solution (7 lts)

2.8 ltrs are alcohol

4.2 ltrs are water

$\therefore$ Ratio of water to alcohol

= 4.2: 2.8   = 3.2

Alternately

(c)   5 lts.   W.A   = 68:32   = 17:8

$\therefore$ Water   = 17/25 x 5   = 17/5

Alcohol   = 8/25 x 5   = 8/5

In 2 lts

Water   = 40/100 x 2   = 4/5

Alcohol   = 60/100 x 2   = 6/5

Final solution

Water alcohol

$^{17}/_5$ + 4/5: 8/5 + 6/5

$^{21}/_5$: 14/5

21: 14

= 3: 2

7.   (a) % Profit taxes and insurance

$^{40}/_{100}$ x 75/100

Amount shared

$$= \frac{100 - (25 + 30) \times 225000}{100}$$

$$\frac{45}{100} \times 225000$$

$$= 101250$$

Amount Cherop received more than Asha:

Ratio of contribution

60,000: 85000: 105 000

12   : 17   : 21

$$\frac{21 - 12}{50} \times 101250 = 18225$$

(b) Profit during 2nd year

$$225000 \times 10/9 = 250, 000$$

Nangila's new ratio

$$= \frac{110000}{275000} = \frac{2}{5}$$

$\therefore$ Nangila's new share of profit

$$= {}^2/_3 \times 112500 = 45000$$

8. $2 \, {}^{11}/{}_{12}$ hours

9.   10 days

10.   Kshs 52

CHAPTER FOUR: MEASUREMENTS I

1.   (a)   (i) $(0.8 \times 1.2) + (1.2) \times 2 + (0.8 \times 1) + \frac{1}{2} \times 0.8 \times 0.3 \times 2$

                             $= 0.96 + 2.4 + 1.6 + 0.24$

                             $= 5.2 \text{ m}^2$

                (ii) $0.6 \times 1.2 \times 2$

                             $= 1.44$

     (b) 300×1.44

                $432 + 1820$

                $= \text{Kshs } 2252$

     (c) $432 (1.5)^2$

                $= \text{Kshs } 972$

2.   (a) $29 + {}^{28}/_2 = 43$

                $= 43 \text{ cm}^2$

     (b) $43.1075 \times 10^4 \times 10^4$

                $1:25 \times 10^8$

                $1:5 \times 10^4$

                $= 1 : 50000$

3.   Area of rectangle $= 19.5 \times 16.5$

         $= 321.75 \text{ cm}^2$

         Area of 4 triangles $= \frac{1}{2} \times 6 \times 4.5 \times 4$

$= 54 \text{ cm}^2$

Area of Octagon $= 321.75 - 54$

4. $V_1 = \pi h (11/2)^2$

$= 3.142 \times (5.5)^2 \times 600$

$V_2 = \pi(9/2)^2 h$

$= 3.142 \times (4.5)^2 \times 600$

Volume of material used $= V_1 - V_2$

$3.142 \times 600 (5.5^2 - 4.5^2)$

$3.142 \times 600 \ (5.5 + 4.5)(5.5 - 4.5)$

$3.142 \times 600 \ (10) (1)$

5.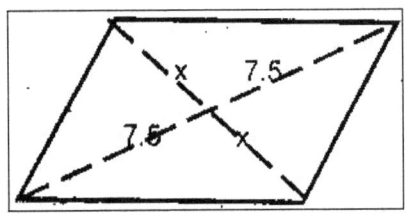

¼ of area $= ¼ \times 60$

$= 15 \text{ cm}^2$

$\therefore ½ \times 7.5 \times X = 15$

$75 x = 30/75 = 4$

$\therefore$ One of the sides $= 7.5^2 + 4^2$

$= 8.5 \text{ cm}$

Perimeter $= 8.5 \times 4$

$= 34 \text{ cm}$

6.   Curved S.A $= ½ \times {}^{22}/_7 \times 2 \times 4.2 \times 150$

= 22 x 0.6 x 150

= 1980 cm²

Area of two semi circular ends = ½ πr² x 2 = 55.44 cm²

Area of rectangular surface = 8.4 x 150

= 1260 cm²

Total surface area = 1980 + 55.44 + 1260

= 3295 . 44 cm²

7.   a)

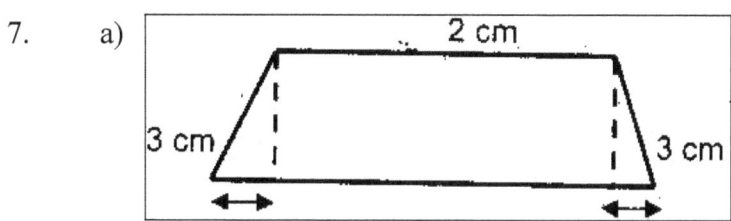

V = cross section area x height

= ½ x 2.4 x (2 + 5.6) x 8

= 72.96 cm³

(b)   Mass = 72. 96 x 5.75 = 419.52g

(c)   (i)   246.24 cross section Area x 8

Cross section Area = $\underline{246. 46}$ = 30.85 cm²

(ii) $\underline{419. 52}$ = $\underline{2}$

185

$$\frac{M_2}{2} \qquad \frac{5}{}$$

$$M_2 = \frac{419.52 \times 5}{2}$$

$$= 1048.8 \text{ g}$$

$$\text{Density} = \frac{1048.8}{246.24} = 4.26\text{g cm}^{-3}$$

8.   Volume of plate $= \dfrac{1.05 \times 1000}{8.4}$

$$= 125 \text{ cm}^3$$

Length of the side $= \dfrac{125}{0.2}$

$$= 25 \text{ cm}$$

9.   (a) L.S.F $= \sqrt{\dfrac{20}{45}}$ or $\sqrt{\dfrac{4}{9}}$ or $\dfrac{2}{3}$

$$\therefore \text{V.S.F} = \frac{8}{27}$$

Capacity of smaller container

$$= \frac{8}{27} \times 0.0945 = 0.28\text{L}$$

(b)   Let depth be h

186

$45 (13 - h) = 20h$

$585 = 65h$

$H = 9$

(c)   Amount in smaller container

$$\frac{1}{5} \times 9 \times 45 + 20 \times 9$$

$= 261$

Height in smaller container

$$\frac{261}{20} = 13.05 \text{ cm}$$

Difference $13.05 - \dfrac{4}{5} \times 9$

$= 13.05 - 7.2$

$= 5.85$

10.   72

11.   (a) 107,800 litres

    (b) 486 days

    (c) 485 days

12. (a) (i) 20.25m^2

   (ii) 50625 kg

   (iii) 5625 kg

   (b) 112.5 (113)

   (c) 4 lorries

13. 1.5 m

14. (a) R = 8.5

   R = 5.5

   V = 1848 cm

15. 97.43 cm^3

16. 267/75 cm^2

17. 270 cm^2

18. 425 ha

CHAPTER FIVE: LINEAR EQUATIONS

1. $3S + 2T = 840$

    $4S + 5T = 1680$

    $12S + 8T = 3360$

    $\underline{12S + 15T = 5040}$

    $7T = 1680$

    $T = 240,\ S = 120$

2. $Y = 2x - 3$

    $X^2 - x(2x - 3) = 4$

    $X^2 - 3x - 4 = 0$

    $(x + 1)(x - 4) = 0$

    $X = -1$ or $x = 4$

    And

3. $5s + 3b = 1750$ ….(i)

    $3s + b = 850$ ….(ii)

    $5s + 3b = 1750$ ….(iii)

    $9s + 3b = 2550$ ….(iv)

    $4s = 800$

    $S = 200$

    $B = 250$

4. Let the cost be Kshs c- cups

S – spoon

$3c + 4s = 324$

$5c - 2s = 228$

$15c + 20s = 1620$

$15c - 6c = 684$

$26s = 936$

$S = 36 \quad c = 60$

5. Let no of ten shillings coin be 6

No of five shilling coin = 2t

No of one shilling coin = 21- 3t

Value = 1 ot + 2t x 5 + (21 – 3t) x 1 = 72

$17t = 51$

$T = 3$

6. $6a + 4b = 7.2$

$2a + 3b = 3.4$

$6a + 4b = 7.2$

$6a + 9b = 10.2$

$5b = 3b = 0.6 \quad a = 0.5$

7. $4p + 6b = 66$ ....(i)

$2p + 5b = 51$ ....(ii)

(a) $4p + 6b = 66$ ….(iii)

$4p + 10b = 102$ ….(iv)

$4b = 36$

$b = 9$ $p = 3$

(b)   Let number of pencils bought be x;

$3x + 9 (x+4) = 228$

$12x = 192$

$X = 16$

8. $x (9x+4) = 32$

$X^2 + 4x – 32 = 0$

$(x – 4) (x + 8) = 0$

$X = 4$ or $x = -8$

Length of room is $4 + 4 = 8m$

9. $2p + 3b = 78$ …………..(i) x 3

$3p + 4b = 108$ …………..(ii) x 2)

$6p + 9b = 234$

$6p + 8b = 216$

$B = 18$

Substituting for b in e.g. ii

$3p + 72 = 108$

$3p = 36$

$P = 12$

10. $m + 14 = 2(s+14)$

$(m + 4) + (s - 4) = 30$

$M = 2s + 14$

$M + s = 38$

$\therefore 2s + 14 + s = 38$

$S = 8$

$M = 30$

$\therefore$ Mother's age when son was born

$= 30 - 8 = 22$

Present 14 years

11.   Ali's age is 16 years

Juma's age is 42 years

12. $s = 30$, $t = 150$ total 180

13.   1080

CHAPTER SIX: COMMERCIAL ARITHMETIC I

1. $25000 - 3750 = 21250$

   Amount to pay $= 21250 + 21250 \times \dfrac{40 \times 2}{100}$

   $= 38250$

   One installment $= \dfrac{38250}{24}$

   $= 1,593.75$

2. (a) $21000 \times 48 - 560,000$

   $1008000 - 560000$

   $= 448,000$

   (b) $448,000 = \dfrac{560,000 \times R \times \$}{100}$

   $R = \dfrac{448,000 \times 100}{560,000 \times 4}$

3. $17500 \times {}^{95}/_{5}$

   $= Kshs\ 322,500$

   Let pineapples sold at Kshs 72 for every 3 be x and at Kshs 60 for every 2 be 144 – x

   $\dfrac{144-x}{2} \times 60 + \dfrac{x}{3} \times 72 = 3960$

   $4320 - 30x + 24x = 3960$

   $6x = 360$

   $X = 60$

4. (a) C.P = 4000 x 100 = 1 2/3% or 5/3 %

    (b) Commission = 5/300 x 98/ 100 x 360,000

        = 5, 880

5. Let the buying price be x

    Profit = (1048 – x)

    Loss – (x – 880)

    4x = 3680

    X = Kshs 920

6. Commission = $\underline{2.4}$ x 100,000 + $\underline{3.9}$ x 180, 000

                        100   100

    = 2,400 + 7, 020

    = Kshs 9, 420

7. Korir   Wangari   Hassan

    ¼ x $^2/_5$ x ¾ x or $^3/_{10}$x $^3/_2$ x ¼ x or $^3/_8$ x

    Bank = x – (¼ x + $^3/_{10}$ x + $^3/_8$x)

    = $^3/_{40}$x

    $^3/_8$ x – $^3/_{40}$ x = 60,000

    X = 200,000

8. (a) Swiss Francs

    52 = 40.63

    1.28

(b) Kshs 40. 63 x 45 . 21

= 1837

9.   Selling price = $\frac{97.5}{100}$ x 120,000

= 117, 000

Commission = $\frac{5}{100}$ x 117, 000

Kshs 5850

Total earning = 5850 + 9000

Kshs 14, 850

10.   105, 000 x 9.74

= Kshs 1, 022, 700

Amt. Remaining = 1, 022, 700 - 403879

= 618, 821

= S.A and Received = 51, 100

11. $\frac{2950000}{118}$

= US dollar 25000

Duty Paid = 25000 x $\frac{20}{100}$ x 76

= Kshs 380, 000

12.   (a)   (i) Kshs 12, 000

          (ii) Kshs 6150

(b)   Kshs 1500 and Kshs 2250

13.   £ 10 or £ 10.6

14.   55086

15.   (a) Kshs 150, 000

(b) Kshs 2025

16.   Kshs 15818.40

17.   11109 or 11110 (table)

18.   Kshs 505, 000

19.   n = 60

CHAPTER SEVEN: GEOMETRY

1. AB correctly constructed

    ABP correctly constructed

    (i)   AD = 4.5 ± 0.1 cm

           Distance A to D = 4.5 x 10 = 45 km

    (ii)   Bearing D from B = 241 ± 1

    (iii) Bearing p from D = 123 ± 2

    (iv)   DP = 12.9 + 0.2 cm

           Distance D to P = 12.9 x 10 = 129 km

2. Location of T

    Location of K

    Location of G

    (a)   Distance TK = 80 ± 2km

           Bearing of T from K: $043^0 \pm 1$

    (b)   Distance GT = 72 ± 2k

           Bearing of G from T: $245^0 \pm 2^0$

    (c)   Bearing of R from G: $130^0 \pm$

3. (a)   Bearing of 060^0 drawn

           Bearing of 210^0 drawn

           Distance on scale drawing

           Representing 150 km

           Representing 1800 km

    (b)   (i)   Actual distance

$(16 \pm 0.1) \times 200$ or equivalent

$= 3200$ Km

    (ii)   Bearing of T from S

           $= 224 \pm 1^0$

    (iii)   Bearing of S from T

           $044^0 \pm 1^0$

.   Measure AB $= 15$ m

Measure 30^0 at B

Construct 90^0 at A

    (a)   Measure height AT $= 105.5 \pm 1$

           Measure height AH $= 8.7 \pm 14$

           Measure height HT $= 1.8 \pm 1$

5. $2n - 4$ right angles

    $2 \times g - 4 = 14$ right angles

    $14 \times 90 = 1260^0$

6. $\sin \beta = \sin 30^0$

    12   15

    $\sin \beta = \dfrac{0.5 \times 12}{15} = 0.4$

    B $= 23.58^0$ (23^035)

    A $180^0 - (30 - 23.58)$

    $= 126.42^0$ $(126^0 25)$

Bearing of Z from X

$180^0 + 126.42^0$

$= 306.42 \ (306^0 \ 25)$

$N = 53^0 \ 25W$

7.   (a) $RA = \underline{30}$ or $RA = 30 \tan 64^0$

               $\tan 26^0$

          $= \underline{30}$ or 30×2.050

          0.4877

          $= 61.51 \ (61.5)$

          $RB = \underline{\ \ 30\ \ }$ or $= 30 \tan s \ 58$

               $\tan 32$

          $= \underline{30}$ or 30×2.050

          0.6249

          $= 48.01 \ (48)$

          $AB \ \sqrt{61.52^2 + 48.01^2}$

          $= \sqrt{3783 + 2305}$ $= 6088$

              $= 78.03$

    (b) $\tan \theta = \underline{48.01}$

               61.51

          $= 0.7805$

          $\theta = 37^0 \ 58$

$$= 322^02 \ (322.03)$$

8.   H = 12 sin 60

       = 10.39

       AD = (12 cos 60) x 2 + 4)

       = 16

       Area [½ x (4 + 16) 10. 39]^2

       = 103.9 x 2

       = 207. 8 cm^2

9.   (a)

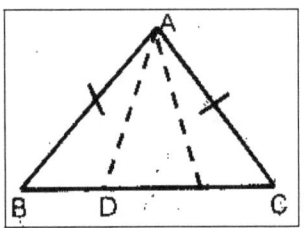

10.   (a)

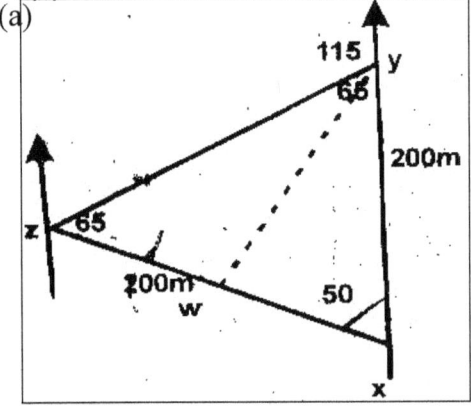

$yz = 200^2 + 200^2 - 2x\ (\ 200\ x\ 200)\ \cos 50$

$yz = 103.\ 53$

Bearing of z from y = 245^0

(b)　(i)

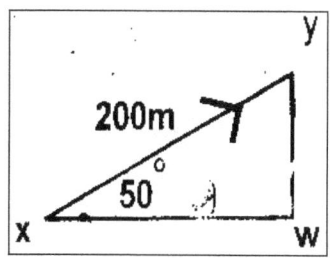

$Yw = 200 \cos 50$

$= 128.6$

(c)　(i)

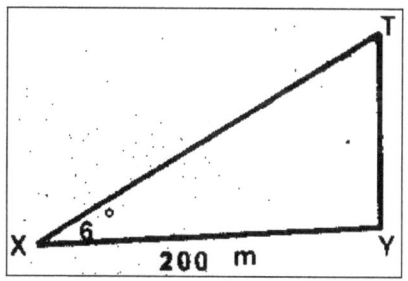

TY = 200 tan 6^0

= 21.02 m

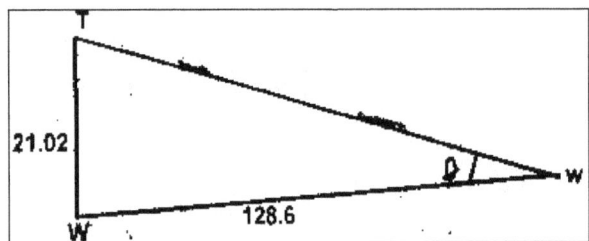

Tan Θ = 21. 02

128.6

Tan Θ = w 0.1635

Θ = 9.28^0

11. (a) From △ BCD

Sin 30^0 = BD

12

BD = 12 sin 30

= 12 x ½

= 6 cm

(b) From △ ABD

$\underline{\text{Sin 45}}$ = $\underline{\text{sin} \angle \text{ADB}}$

6   8

Sin ∠ ADB = $\underline{\text{8 sin 45}}$

6

$$= \underline{4 \times 0.7071}$$

$$3$$

$$= 0.9428$$

$$\angle\, ADB = 70.53$$

12. (a) $\angle\, ADE = 36^0$

    (b) $\angle\, AEF = 66^0$

    (c) $\angle\, DAF = 12^0$

13.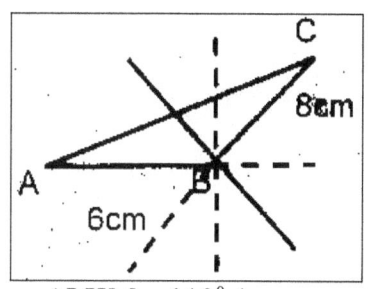

14. $\angle\, LKM = 110^0$ (- seen or implied)

    $\angle\, KLM = 35^0$ (or kml = 35^0)

    Bearing is 185^0

15. (a) Diagram

    (b) (i) 73 ± 1 km

        (ii) $102^0 \pm 1^0$ or 578^0 E $\pm 1^0$

16. (a) Diagram

        600 km am 500 km seen or used

        Scale used

        Bearing and distance of P

        Bearing and distance of Q

(b) 1060 ± 10 km

(c)   (i) 254 ± 1^0

(ii) 0.74 ± 1^0

17.   (i)   45 km

(ii) 124 ± 1

(iii) 123 ± 2

(iv)   129 km

18.   Location of T

Location of K

Location of G

(a)   Distance TK = 80 ± 2 km

Bearing of T from K: $043^0 \pm 1$

(b)   Distance GT = 72 ± 2k

Bearing of G from T: $245^0 \pm 2^0$

(c)   Bearing of R from G: $130^0 \pm 2^0$

19.   (a) $< BAE = \dfrac{540^0}{5} = 108^0$

(b) $< BED = 108^0 - 36^0$

$= 72^0$

(c) $<BNM = 90^0 - 36^0$

$= 54^0$

20.

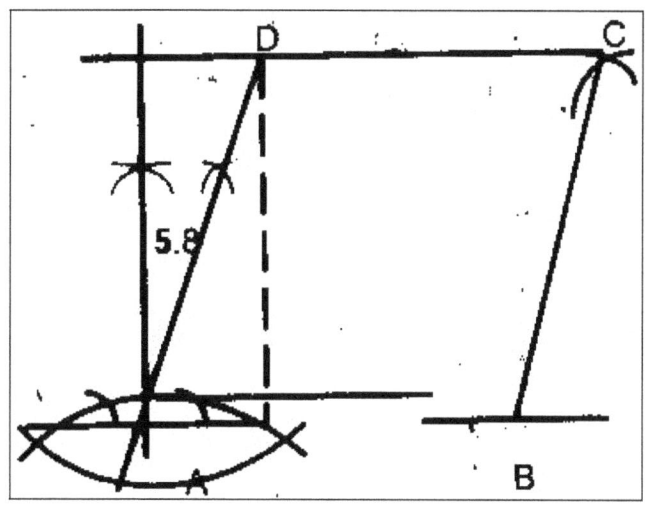

21. $2x + \frac{1}{2}x + x + 40 + 100 + 130 + 160 = 720$

$$\frac{7x}{2} = 280$$

$X = \frac{280 \times 2}{7} = 80^0$

Smallest angle $\frac{1}{2}x = 40^0$

22. Ext angle $= 180 - 156$

$= 24$

$N = \frac{360}{24}$

$= 15$

CHAPTER EIGHT: COMMON SOLIDS

1.   (a)

    (b)   Four (4) planes of symmetry

2.

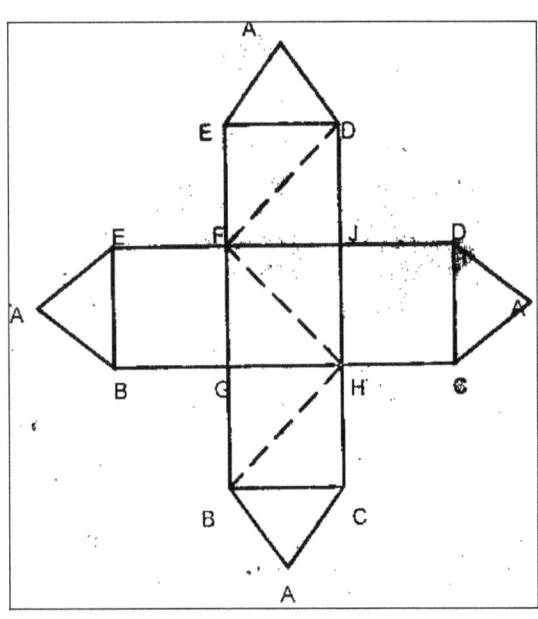

3.

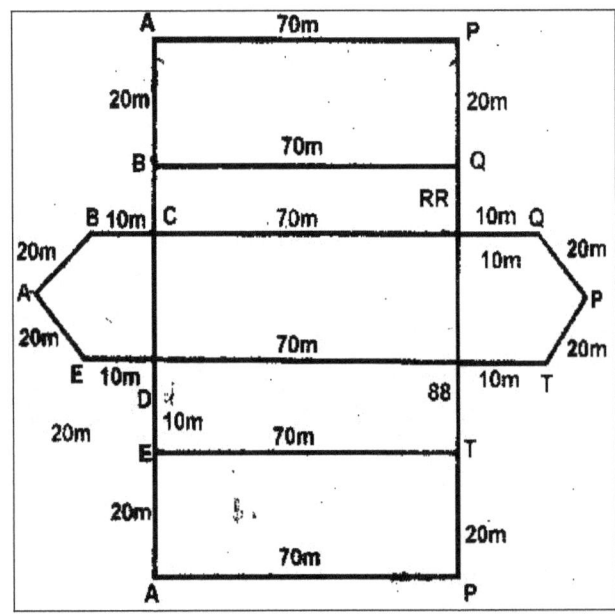

4.	(a)

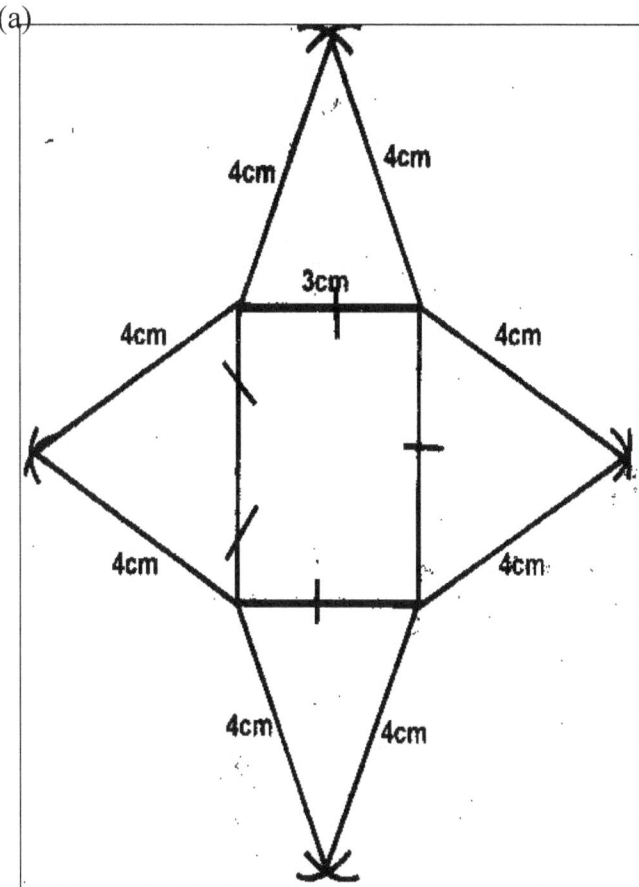

4cm 4cm

3cm

4cm 4cm

4cm 4cm

4cm 4cm

	(b)	VO = 3.7 cm	(Not to scale)

5.

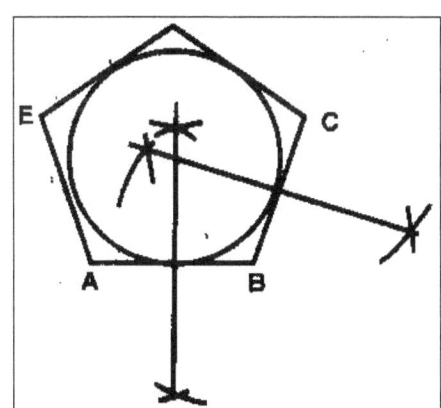

E	C

A	B

6.

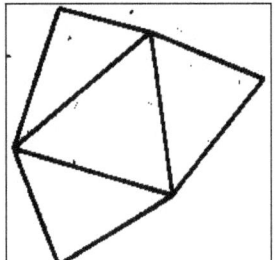

7. (a)

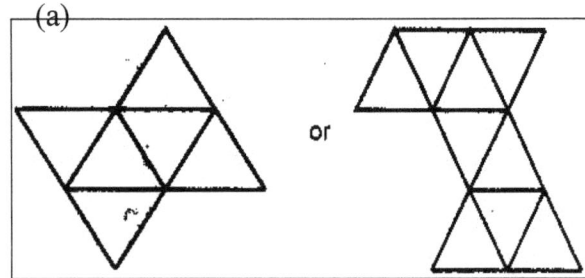

or

(b)   64.95 cm²

8.

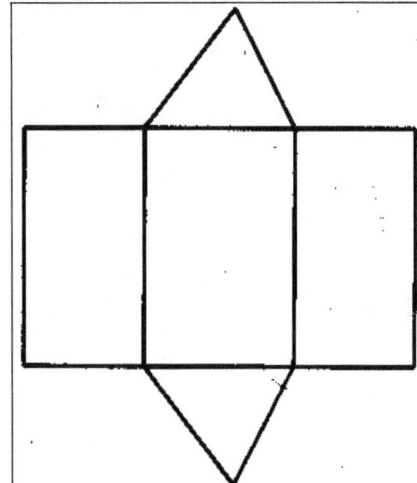

208

NINE: NUMBERS II

1.

No	Log
36.15	1.5581
0.02573	2.4104
	1.9685
1.838	0.2874
	1.6811 ÷ 3
	[3 + 2.6811] ÷ 3

7.829 x 10^{-1} 1.8937

= 0.7829 or 7828

2. $2^{4(x2)} = 2^{3(4-3)}$

$4x^2 = 12x - 9$

$4x^2 - 12x + 9 = 0$

$= (2x - 3)(2x + 9) = 0$

$X = 3/2$ or 1.5

3.

No	Log
(1934)2	3.2865 x 2
	6.5730 ←
0.0034	3.5105 ÷ 2

$$\frac{4 + 1.5105}{2}$$

$$\underline{2.75525}$$

5. 32825

| 436 | 2.63950 |
| 4.884 x 10² | 2.6888 |

= 488.4 or 488.5

4. | No | Log |
|---|---|
| 55.9 | 1.7474 |
| 0.2621 | 1.4185 |
| 0.01177 | 2.0708 |

= 3.4893

$$\frac{5 + 2.4893}{5} \quad = \underline{1.4979}$$
$$2.2495$$

1.776 x 10²

= 177.6

5. 2² x 5²ˣ

(2 x 5)^(2x − 2) = 10

6. | No | Log |
|---|---|
| 3.256 | 0.5127 |
| 0.0536 | 2.7292 |

1.2419 ÷ 3

$(3 + 2.2.2419) \div 3$

0.5589   1.7473

7. $2^{5(x-3)} \times 2^{2(x+4)} = 2^6 \div 2^x$

5(x-3) + 3(x + 4) = 6 – x

9x = 9

X =1

8. $(3^4)^{\,2x} \times (3^3)^x = 3^6$

8x + 3x – 2x = 6

9x = 6

X = 2/3

9. $\underline{1\quad\;\;} = 0.04072$

24.56

4.3462 = 18.89

0.04072 + 18.89 = 18.93072

= 18.93

10. | No | Log |
|---|---|
| 0.032 | 2.5051 |
| 14.26 | 1.1541 |
| | 1.6592 |
| 0.006 | 3.7782 |
| | 1.8810x $^2/_3$ |
| 17.95^4 | 1.2540 |
| | 17.95 |

11. $x = \frac{1}{2}$

12. 0.01341

13. $m = -3$

14. $y = 0$

15. 177.6

16. 2.721

17. 0.0523

18. 0.001977

CHAPTER TEN: EQUATIONS OF LINES

1. (a) $OT = \frac{1}{3} \begin{pmatrix} -\frac{1}{2} \end{pmatrix} + \frac{2}{3} \begin{pmatrix} \end{pmatrix} \begin{pmatrix} \frac{4}{10} \end{pmatrix} = \begin{pmatrix} \frac{3}{6} \end{pmatrix}$

    (b) (i) Gradient PQ = 4

            Gradient normal / $\perp$ = - ¼

         (ii) $y - 6 = -1$

            $X - 3$

            $4(y-6) = -1 \ (x-3)$

            $4y - 24 = -x + 3$

            $4y = -x + 27$

         (iii) $\sqrt{(6\frac{3}{4} - 6)^2 + (3-0)^2}$

            $= \sqrt{9.5625}$

            $= 3.092$

            $= 3.09 \ (3 \ s.f)$

2. $L_1 \ \dfrac{-2 = 5}{x-1}$

    $y = 5x - 3$

    L_2 at $x = 4$, $y = 17$

$$\underline{y-17} = \underline{-1}$$
$$x-5 \qquad 5$$

$$y = {}^1/_5\,x + {}^{89}/_5$$

3.   Midpoint of PQ $= \dfrac{5 + (-1)}{2} - \dfrac{(4 + (-2)}{2}$

$$= 2, -3$$

Gradient of PQ $= \dfrac{-4 - (-2)}{5 - (-1)}$

$$= -1/3$$

$\therefore$ Gradient of $\perp$ bisector $= 3$

Equation of $\perp$ bisector $= \dfrac{y - (-3)}{x - 2} = 3$

$$y + 3 = 3x - 6$$

$$y = 3x - 9$$

4. $7y = 3x - 30$

$$Y = \dfrac{3x}{7} - \dfrac{30}{7}$$

Y intercept $= \dfrac{-30}{7}$

X intercept $= 10$

A is $(10, 0)$

Based on line $y = -x$

$$Y = \underline{3x} - \underline{30} = \underline{3(-y)} - \underline{30}$$

$$7 \quad 7 \quad 7 \quad 7$$

$$Y = -3y - 30$$

$$7 \quad 7$$

$$^{10y}/_7 = ^{-30}/_7$$

$$Y = -3$$

$$\therefore x = 3$$

B (3, -3)

5. $\underline{8-k} = -3$

      k-3

      8-k = -3k + 9

      2k = 1

      $\therefore k = \frac{1}{2}$

      Taking a general point (x, y)

      $\underline{Y - 8} = -3$

      X - ½

      y- 8 = -3x + $^3/_2$

      3x + y = 9 ½ or 6x + 2x + 2y = 19

6. $\underline{6+2}$ $\underline{1+3}$ = (4,2)

      2   2

      $\underline{1 -3}$ x u$_2$ = -1   (M$_2$ = 2

      6 – 2

      Y – 2 = 2

      X – 4

$\therefore 2x - y = 6$

7.   (a) $\frac{1}{5}$

      (b) $y = -5x + 7$

8. $y = 2x - 3$

9. $y = -2x + 13$

10. $y = \frac{2}{5}x + 5$

11.   Gradient $= \frac{4}{3}$ or $1\frac{1}{3}$

      Y – intercept $= -3$

CHAPTER ELEVEN: TRANSFORMATIONS

1.
$$\begin{pmatrix} x \\ y \end{pmatrix} = \begin{pmatrix} -1 \\ 2 \end{pmatrix} - \begin{pmatrix} 1 \\ 2 \end{pmatrix}\begin{pmatrix} -2 \\ 0 \end{pmatrix} =$$

x^1	=	-3	+	-2	-5
y^1		-3		0	-3

$$\Rightarrow (x', y') = (-5, -3)$$

2.

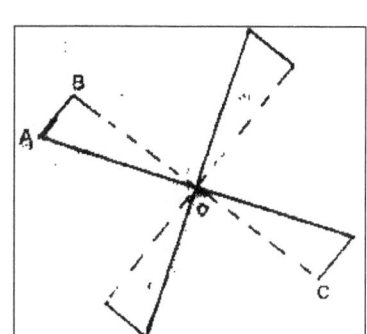

217

3. $\begin{pmatrix} -5 \\ 4 \end{pmatrix} + T = \begin{pmatrix} -1 \\ -1 \end{pmatrix}$

$T = \begin{pmatrix} -1 \\ -1 \end{pmatrix} - \begin{pmatrix} -5 \\ 4 \end{pmatrix}$

$T = \begin{pmatrix} 4 \\ -5 \end{pmatrix}$

$\begin{pmatrix} -4 \\ 5 \end{pmatrix} + \begin{pmatrix} 4 \\ -5 \end{pmatrix} = \begin{pmatrix} 0 \\ 0 \end{pmatrix}$

4. $\begin{pmatrix} 0 & 1 \\ -1 & 0 \end{pmatrix}\begin{pmatrix} 2 & 4 & 1 \\ 1 & 1 & 6 \end{pmatrix} = \begin{pmatrix} 1 & 1 & 6 \\ -2 & -4 & -1 \end{pmatrix}$

5. (a) (i) Diagram

   (ii) A" (2) B" (7 -2) C"(5, -4) D" (3, -4)

   (b) A" (2) B" (-7, -2) C" (-5, -4) D"(-3, 4)

   (c) Half turn

      Centre (0,0)

6. $\begin{bmatrix}5\\4\end{bmatrix} - \begin{bmatrix}3\\2\end{bmatrix} = \begin{bmatrix}2\\6\end{bmatrix}$

OQ $= 2 + 2 = 4$

$\begin{bmatrix}5\end{bmatrix}\begin{bmatrix}-6\end{bmatrix}\begin{bmatrix}\ \end{bmatrix}$

$\sqrt{}$

PQ $= 4 \quad 5 \quad = \quad -1$

$\quad\quad -1 \quad -4 \quad\quad 3$

PQ $= \quad (-1)^2 + 3^2$

$= \sqrt{10}$

7. (a) Translation $= 10 \quad - \quad -2 \quad = \quad 12$

$\quad\quad\quad\quad\quad\quad 10 \quad\quad 3 \quad\quad\quad 7$

$\therefore Q = \begin{bmatrix}1\\3\end{bmatrix} + \begin{bmatrix}12\\7\end{bmatrix} = \begin{bmatrix}13\\10\end{bmatrix}$

$= 13, 10$

(b) $\quad$ m $\quad$ -2m $\quad$ - n $\quad$ 1 $\quad$ = $\quad$ -12

$\quad\quad\quad\quad\quad$ 3m $\quad\quad\quad\quad$ 3 $\quad\quad\quad\quad\quad$ 9

$\quad\quad$ -2m $\quad$ - $\quad\quad$ n $\quad\quad$ = -12

$\quad\quad$ 3 $\quad\quad\quad\quad$ 3n $\quad\quad$ 9

$-2m - n = -12$ ………….. x3

$3m - 3n = 19$ ………….. x1

$-6m - 3n = 36$

$3m - 3n = 9$

$-9 = -45$

M = 5; n = 2

8. $\quad$ (a) $\quad$ Reflection along y- axis $\quad\quad$ (x =0)

$\quad\quad$ (b) $\quad$ (on graph)

$\quad\quad$ (c) $\quad$ Rotation about (0,0) through 90^0

$\quad\quad$ (d) $\quad$ On the graph

$\quad\quad$ (e) $\quad$ P" Q" R" and P" Q" R"

$\quad\quad\quad\quad$ P Q R and P' Q' R'

$\quad\quad\quad\quad$ P" Q' R' and P" Q" R"

CHAPTER TWELVE: MEASUREMENTS II

1. $\frac{4}{3} \times \frac{22}{7} \times r^3 = \frac{1}{3} \times \frac{22}{7} \times 9 \times 9 \times 12$

    $R^3 = 243$

    $R = 6.24$ or equivalent

    $A = 4 \pi r^2 = 4 \times \frac{22}{7} \times 6.24 \times 6.24$

    $= 489.5 \text{ cm}^3$

2. (a) Area of path $= \frac{22}{7} \times 49^2 - \frac{22}{7} \times 35^2 = 36976 \text{m}^2$

    Area of slab=

    $\frac{22}{7} \times 35 - 4 \times 4 \times 3 = 3850 - -48 = 3802 \text{m}^2$

    Total cost $= 3696 \times 300 + 3802 \times 400 = 2629600$

    Amount nit spent $= \frac{20}{100} \times \frac{115}{100} \times 2629600$

    Kshs 604808

    (b) Actual expenditure

    $= \frac{80}{100} \times \frac{115}{100} \times 2629100 = 2419232$

3. $1 + x^2 = (2x -1)^2 -1$

    $3x^2 - 4x - 1 = 0$

    $X = 1.549 \text{m}$

4. Volume of the cone $= \frac{1}{3} \times \frac{22}{7} \times 7 \times 7 \times 18$

    $= 924 \text{ cm}^2$

    Let change in height be H

    Volume of water displaced $= 22/7 \times 14 \times 14 \times H$

    $= 616 \text{ cm}^3$

    $\Pi \times 14 \times 14 \times H = \frac{1}{3} \pi \times 7 \times 7 \times 18$

$$H = \frac{49 \times 6}{14 \times 14} = 1.5 \text{ cm}$$

5. (a) $\frac{1}{3} \pi \times r^2 \times 9 = 270 \pi$

$$R^2 = \frac{270 \times 3}{9} = 90$$

$$R = \sqrt{90} = 9.49$$

6. Initial volume $= \frac{4}{3} \pi \times 2^3$

$$= \frac{32}{3} \pi$$

New vol. $= \frac{32}{3} \pi \times \frac{337.5}{100}$

$$= 36 \pi$$

7. (a) $y^2 - (\frac{1}{2} \times x^2 \times 4)$

$$Y^2 - 2x^2$$

(b) $2x^2 = 14^2$

$$X = 7 \sqrt{2}$$

(c) Area of the octagon

$$Y = 14 + 2x = 14 + 2 \times 9.9$$

$$A = y^2 - 2x^2$$

$$= (3.38)^2 - 2 \times (9.9)^2$$

$$= 946.44 \text{ cm}^2$$

8. Volume $= \frac{1}{3} \times 12 \times 9 \times 6$

$$= 216 \text{ cm}^3$$

9. (a) (i) $A = \frac{22}{7} \times 4.2 \times 4.2 = 55.44$

= 55.44 cm^2

(ii)   Let slanting length cone be L

$\therefore \dfrac{L-8}{L} = \dfrac{3.5}{4.2}$ or equivalent

L= 48 cm

Curved area of frustum

= 22/2 (4.2 x 48- 3.5 x 40)

= 193.6 cm^2

(iii)   Hemispherical surface area

= ½ x 4 x 22/7 x 3.5 x 3.5

= 77 cm^2

(b)   Ratio of area = 81.51: 326. 04

= 1.4

Ratio of lengths = 1.2

Radius of base = $\dfrac{4.2}{2}$

= 2.1 cm

10.   ½ x 5 x 5 sin 120

½ x 25 x 0.8666

10.83 cm^2

11.   BO – OD = 15^2 – 12^2 = 81 = 9

Area = ½ x 9 x 12 x 2 x ½ x 9 x 18 x 2

= 108 + 162

270 cm^2

12. $\frac{1}{3}$ x $\frac{22}{7}$ x 6 x 6 x 9 + $\frac{1}{2}$ x $\frac{4}{3}$ x $\frac{22}{7}$ x 6 x 6 x 6

339.4 + 452.6

= 792

13. X –section Area = $\frac{22}{7}$ ($4^2 - 3^2$) cm^2

= 7 x $\frac{22}{7}$

Vol = 7 x 0.2 x $\frac{22}{7}$

= 4.4 cm^3

14. Let the width be x

($\frac{3}{2}$ x + x) 2 + 2x= 21

3x + 2x + 2x = 21

7x = 21

X = 3 cm

15. $V_1 = \pi$ h $(11/2)^2$

= 3.142 x $(5.5)^2$ x 600

$V_2 = \pi$ $(9/2)^2$ h

= 3.142 x $(4.5)^2$ x 600

Volume of material used = $V_1 - V_2$

3.142 x 600 ($5.5^2 - 4.5^2$)

3.142 x 600 (5.5 + 4.5) (5.5 – 4.5)

3.142 x 600 (10) (1)

= 18.852 cm^3

16.   X- Section Area = (½ x 5 x 5 sin 60) x 6

        = 10.825 x 6

        = 64.95

        Volume = 64.95 x 20

        1,299 cm^3

17.   Curved S.A = ½ x $^{22}/_7$ x 2 x 4.2 x 150

        = 22 x 0.6 x 150

        = 1980 cm^2

        Area of two semi- circular ends = ½ πr^2 x 2

        = 55. 44 cm^2

        Area of rectangular surface = 8.4 x 150

        = 1260 cm^2

        Total surface area = 1980 + 55. 44 + 1260

        = 3295. 44 cm^2

18 (a)

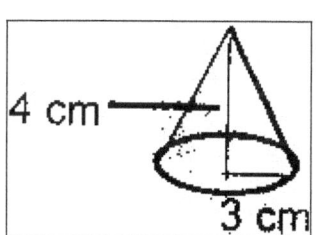

        A_c = $\pi r l$

                = 3.142 x 3 x 5

                = 47.13 cm^2

A_{cs} $= \pi Dh$

$= 3.142 \times 6 \times 8$

$= 150.82 \text{ cm}^2$

As $= \frac{1}{2} 4\pi r^2 = 2\pi r^2$

$= 2 \times 3.142 \times 9$

$= 56.56 \text{ cm}^2$

Ext S.A $= 47.13 + 150.82 + 56.56 = 254 \text{ cm}^2$

(b) c.s.f $= 15/600$ $= 1/40$

$\therefore$ A.S.F $= 1/1600$

$\dfrac{254.5}{\text{actual area}}$ $= \dfrac{1}{1600}$

Actual Area $=$ $407,200 \text{ cm}^2$

Actual area $=$ 40.72 m^2

$\dfrac{40.72}{20} \times 0.75$ $= 1.527 \text{ ltrs}$

19. (a) Let width of path be xm

L $= 10 + 2x$

W $= 8 + 2x$

$(10 + 2x)(8 + 2x) = 168 \text{m}^2$

$80 + 16x + 20x + 4x^2 = 168$

$4x^2 + 36x - 88 = 0$

$X^2 + ox - 22 = 0$

$(x-2)(x+11)=0$

$\therefore x = 2m$

(b) (i)   Area of path $= 168 - (10 \times 8) = 88m^2$

Area covered by corner slabs

$= 4(2x) = 16m^2$

Area to be covered by smaller slabs

$= 88 - 16 = 72m^2$

No. of smaller slabs used

$= \dfrac{72 \times 100 \times 100}{50 \times 50} = 288$

(ii)   Cost of corner slabs

$600 \times 4 \qquad = 2400$

Cost of smaller slabs

$288 \times 50 = 14400$

Total cost $= 2400 + 14400$

Kshs 16,800

20. $\cos 0 = 2.5/5 = 0.5$

$\theta = 60^0$

Surface under water $= \dfrac{2 \times 60}{360} \times \pi \times 10 \times 12 = 125.7$

21.   Area of each sector

$\dfrac{60}{360} \times \pi \times 6^2$

= 18.84955592

Area of Δ = ½ x 6 x 6 x sin 60^0

= 15.5884527

$\therefore$ Area of the shaded region

15.58845727 + 2(18.84955592) – 15.5884527)

= 15.58845727 + 6.522197303

= 22. 11065457

= 22.11

22. (i) 93.54 cm^2

(ii) 28.06 cm^2

23. (a) 107,800 litres

(b) 486 days

(c) 485 days

24. 72

25. (a) Sketch

(b) 10.44 cm

CHAPTER THIRTEEN: QUADRATIC EXPRESSIONS AND EQUATIONS

1. $\dfrac{2x - 2}{6x^2 - x - 12} \div \dfrac{x - 1}{2x - 3}$

    $\dfrac{2(x - 1)}{3x + 4)(2x - 3)} \times \dfrac{(2x - 3)}{x - 1}$

    $= \dfrac{2}{3x + 4}$

2. $y = 2x - 3$

    $x^2 - x(2x - 3) = 4$

    $x^2 - 3x - 4 = 0$

    $(x + 1)(x - 4) = 0$

    $x = -1$ or $x = 4$

    And

    $y = -5$ or $y = 5$

3. $7^{2(x - 1)} + 7^{2x} = 350$

    $7^{(2x + 2)} + 7^{2x} = 350$

    $49(7^{2x}) + 7^{2x} = 350$

    $7^{2x}(49 + 1) = 350$

    $7^{2x}(50) = 350$

    $7^{2x} = 7;\ 2x = 1$

    $x = \frac{1}{2}$

4. $3x^2 - 1 - (2x + 1)(x-1)$

$X^2 - 1$

$X^2 + x$

$X^2 - 1$

$\dfrac{X(x+1)}{(x+1)(x-1)} = \dfrac{X}{x-1}$

5. $3x^2 - 3xy + xy - y^2$

$3x(x - y) + y(x - y)$

$(x - y)(3x + y)$

7. $(a + b)(a - b)$

$(2557 + 2547)(2557 - 2547)$

5104×10

51040

8. (a) (i) $(x + y)^2 = x^2 + 2xy + y^2 = 3^2$

$\therefore x^2 + 2xy + y^2 = 9$

(ii) $2xy = 9 - (x^2 + y^2)$

$= g - 2g$

$= -20$

(iii) $(x-y)^2 = x^2 + y^2 - 2xy$

$= 2g - (-20)$

$= 49$

(iv) $x-y = \pm\sqrt{49}$

$= +7$ or -7

230

(b)   x + y = 3

$\underline{X - y = 7}$ x + y = 3

2x = 10   x – y = -7

X = 5   2x = -4

Y = -2   x = -2

          Y = 5

9.   (3a + b) (a+ b)

(4a – b) (a+ b)

3a + b

4a – b

10.   (a)   10x + y

(b)   3(x+ y) + 8 = 10x + y ...... (i)

10y + x = 10x + y + 9 ....(ii)

2y – 7x = -8

9y – 9x = 9

18y – 18x = 18

18y – 63x = -72

45x = 90

X = 2; y = 3

Xy = 23

11. $2a^2 - 3ab - 2b^2 = 2a + b)(a - 2b)$

$4a^2 - b^2 = (2a - b) (2a + b)$

(2a+ b) (a – 2b)

(2a – b) (2a + b)

<u>a – 2b</u>

2a - b

12. (3t + 5a) (3t – 5a)

3t + 5a (2t + 3a)

= 3t – 5a

   2t + 3a

13. <u>$p^2 + 2pq + q^2$</u>

$P^3 – pq^2 + p^2q – q^3$

<u>(p + q) (p + q)</u>

$P^2 – q^2$) (p + q)

<u>(p+ q) (p + q)</u>

(p – q) (p + q) (p + q)

<u>   1</u>

(p – q)

14. $14(x^2 – y^2) (x^2 + y^2) (x^4 – y^4)$

= $(x^4 – y^4) (x^4 – y^4)$

= $x^8 – 2x^4 y^4 + y^8$

$(x^2 – y^2) (x^6 – x^2 y^4 + x^4 y^2 – y^6)$

15. $x + y = 40 \Rightarrow y = 40 - x$

$\therefore$ Sum of the squares in terms of x

$S = x^2 + (40 - x)^2$

$2x^2 - 80x + 1600$

16. $\underline{15a^2 b - 10ab^2}$ $= \underline{5ab (3a - 2b)}$

$3a^2 - 5ab + 2b^2$ $(3a - 2b) (a- b)$

$= \underline{5ab}$

a-b

17. (i) $9a + 6$

(ii) 18

18. $x = 3/2$

19. 3

20. $x = 4$

21. $y = 0$

22. $2x + y$

$X - 3y$

23. Juma = 42 years

Ali = 16 years

24. $\underline{x - 8}$

$X - 2$

25. $d = 49$

CHAPTER FOURTEEN: INEQUALITIES

1. $2 \leq 3 - x$ $3 - x < 5$

    $-1 \leq -x$ $-x < 2$

    $1 \geq x$

    $-2 \, x \leq 1$ or $1 \geq x > -2$

2. $4 - 2x < 4x - 9$

    $13 < 6x$

    $^{13}/_6 < x$

    $4x - 9 < + 11$

    $\Rightarrow 3x < 20$

    $X < {}^{20}/_3$

    Integral value of x = (3,4,5,6)

3. $3 - 2x < x$

    $3 - 2x + 2x < x + 2x$

    $3 < 3x$

    $1 < x$

    $x = \leq \underline{2x + 5}$

    $\qquad \qquad 3$

    $3x \leq 2x + 5$

    $x \leq 5$

    $= 1 \leq x \leq 5$

4. $x = \geq -16$

5. (a) $x > 0, y > 0$

    (b) $200x + 1400y \leq 9800$ or $x + 7y \leq 49$

    (c) (i) $x = 10, y = 4$

         (ii) $x = 7, y = 6$

              Distance = 690 km

CHAPTER FIFTEEN: CIRCLES

1. $< PCB = 40$ or $<DCQ = 40$

    Or $< BCD = 140^0$

    $\therefore < BAD = 40^0$

2. (i) $< BAC$ or $< BCA = \frac{1}{2} \times 90 = 45^0$

    $< CAD = 180 - (90 + 25)$

    $\frac{1}{2} \times (180 - 2 \times 25)$

    $= 65^0$

    $<BAD = 45^0 + 65^0 = 110^0$

    (ii) Obtuse $< BOD = 2(45 + 25)$

    $= 140^0$

    $= BGD = 70^0$

    (iii) $< ABC = < BAC = 45^0$ base

    $< ABE = <ACB = 45^0 <$ s is alt- segment

    $< CBF = < BAC = 45^0 <$'s alt- segment

    $\therefore < ABE = CBF$

3. (a) $< QTS = 40^0$

    $<$ s' in alt- segment

    (b) $< QRS = 10^0$

    Reasons: $< SQT = 90^0$ on semi circle

    $\Rightarrow < TSQ = 50^0$

    $\therefore QRS = 50 - 40$ etc $<$ of Δ

(c) $< QVT = 35^0$

Reasons $< QVT = SQV$ alt $< S$

(d) $< UTV = 15^0$

Reasons $< QUT = UTV + QVT$

Ext $<$ of triangle

$\therefore = 50 - 35$

4.   (a) $< RSTY = 104$

(b) $< TSU = 180 - 104 = 76^0$

$< QTS = 180 - (90 + 37) = 53^0$

Or $< QRU = 180 - 48 = 132$

$< SUT = (48 + 53)^0 - 76^0$

Quadrilateral

OR $360 - (132 + 76 + 127)$

$= 25^0$

(c)   Obtuse $< RUT = 76 \times 2$

$= 152^0$

(d) $<PST = 70 - 48$ or equivalent

$= 42^0$

5.   (a)   (i) $<CBD = 90 - 42 = 48$

Subtended by diameter

(ii) $< BOD = 180 - 42 = 138^0$

Cyclic quadrilateral

Reflex BOD $= 360 - 138 = 222^0$

(b)   In △ BAD

< BAD = ½ x 138 = 69^0

< ADB = 180^0 – 42 + ½ x 138)

= 180 – 111

= 69^0

6.   (a)   < ECA = 28^0

< CEG = 120^0 or < EAG = 120^0

< ABC = 88^0

7.   < RST = 35 + 20^0 = 55

55^0

8.

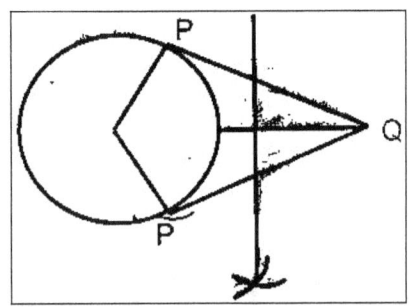

CHAPTER SIXTEEN: LINEAR MOTION

1.   (a) $\underline{300}$

           T – 1

    (b)   Speed of the bus = 500

           T -1

           $\underline{500}$: $\underline{300}$ = 5: 3

           T -1   t – 1

2.   Speed of slower athlete = $\underline{800}$

                                           108

        Distance = $\underline{800 \times 4}$

                    108

    = 29.63

3.   Distance covered by bus A at loan

    = 90 x 2 = 180 km

    Bus B time between 2 stops

    $\underline{72}$ = 1.2 hours

    60

    Bus B leaves L at 9.17 am

    Distance between 9: 17 and 10.00 a.m

    = 60 x 43 = 43 km

        60

    At 10 am Bus B has covered 72 + 43 = 115 km

    Distance between Bus A and B at 10 am

$360 - (180 + 115) = 65$ km

4. Let dist covered by bus be x km

$$\underline{X} = \underline{220} - x + \underline{3}$$
$$60 \quad 80 \quad\quad 4$$

$4x = 3 (220 - x) + 3 \times 60$

$4x = 660 - 3x + 180$

$7x = 840$

$X = 120$

ALT METHOD 2

Let time taken when both are moving to be t hrs

$60 (1 + ¾) = 220 - 80 t$

$\Rightarrow t = 1 ¼$ h

Time bus moving $= 1 ¼ + ¾ = 2$ h

Distance bus covered $= 60 \times 2$

$\quad\quad\quad = 120$

ALT METHOD 3

Relative velocity $= 140$

$\therefore$ time taken $= \dfrac{220 - ¾ \times 60}{140}$

$= 1.25$ h

$\therefore$ Distance bus covered

$1.25 \times 50 + 45 = 120$

5.   (a) $\dfrac{d}{50} - \dfrac{d}{80} = 3$

            $\dfrac{8d - 5d}{400} = 3$

            $3d = 1200$

            $D = 400$ km

    (b)   (i) $400 \times 0.35 + 400 \times 0.3 = 260$ ltr

            (ii)   Total time

                 $\dfrac{400}{50} + \dfrac{400}{80} = 12$ hours

                 Average consumption $= \dfrac{260}{13}$

                           $= 20$ litres/ hr

6.   (a)   Time taken by lorry $= \dfrac{280h}{X}$

            Time taken by car $= \dfrac{280 h}{x + 20}$

            $\dfrac{280}{X} - \dfrac{280}{x + 20} = \dfrac{7}{6}$

            $\dfrac{280 (x + 20) - x (280)}{X (x + 20)} = \dfrac{7}{6}$

241

$$280x + 5600 - 280x = 7/6 (x^2 + 20x)$$

$$7x^2 + 140x - 33600 = 0$$

$$X^2 + 20x - 4800 = 0$$

$$(x + 80)(x - 60) = 0$$

$$X = -80 \text{ or } x = 60$$

$\therefore$ Speed of lorry = 60 km/h

(b)   Speed of car = 80 km/ h

Time taken to meet = 4h

Distance covered by lorry in 4 hours = 60 x 4 = 240 km

Distance covered by car at meeting point = 240 km

Time taken by car = $\dfrac{240}{80}$

= 3 hrs

$\therefore$ Car left M at 9.15 am

7.   Distance covered by bus in 2 ½ hrs

$60 \times \dfrac{5}{2}$ = 150 km

(a)   (i)   500 – 150 = 350 km

(ii)   Overtaking speed = 100 – 60 = 40 km/h^{-1}

Distance = 150 km

Time taken to overtake = $\dfrac{150}{40}$ = 3 ¾ hrs

Distance traveled by car to catch up

$$= 100 \times 15/4 = 375 \text{ km}$$

(b)   Distance remaining = 500 – 375 = 125 km

Time taken by bus to cover 125 km

$$= \frac{125}{60} = 2 \tfrac{1}{2}$$

Time left for the car after rest

= 2 hrs 5 min – 25 min

= 1 hr 40 min

∴ New average speed = 125 ÷ 1 2/3 = 75 kmh^{-1}

8.   Amount of fuel used = $\dfrac{120}{4} \times \dfrac{8}{3}$

Amount of money spent = 80 x 59

= 4720

9.   (a) 15 km

(b) 71.25 km

10.   97

11.   9.20 am

12.   (a) 20 m/s

(b) 220 m

CHAPTER SEVENTEEN: QUADRATIC EXPRESSIONS AND EQUATIONS II

1. a) (i) $b - a = 35$........(i)

    $7b - 490a = 39.9$....(ii)

    $A = 4.9$ $b = 40$

    (ii) $S = 4.9t^2 + 40t + 10$

t	0	1	2	3	4	5	6	7	8	9	10
t	10		70.4	85.9	91.6	87.5	73.6		16.4	-26.4	

(b) (i) Suitable scale

    Plotting

    Curve

    (ii) Tangent at t = 5

    Velocity = -9.0 ± 0.5 m/s

2. (a)

X	-3	-2	-1	0	1	2	3	4
Y	6	0	-4	-6	-6	-4	0	6

(b) Suitable scale

    Plotting

    Curve

(c) $y = -3x - 4$

    Line drawn

3. (a)

X	-4	-3	-2	-1	0	-0.5	1	2	3	4	5
Y	-14	-6	0	4	6	6.25	6	4	0	-6	14

B_1 For all values correct

Line graph = $y = 2 - 2x$

(b) $x = -1$ $x = 4$

(c) $6 + x - x^2 = 2 - 2x$

$X^2 - 3x - 4 = 0$

4. (a)

x	-2	-15	1	0	1	2	3	4	5
X3	8	-3.4	-1	0	1	8	27	64	125
-5x2	-20	-11.3	-5	0	-5	-20	-45	-80	-125
2x	-4	-3	-2	0	2	4	6	8	10
y	9	9	9	9	9	9	9	9	9

(b)   On the graph scale

Plotting

Curve

(c) 2.15 ± 0.1

(d) $y = 4 - 4x$

$X = 0.55 \pm 0.1$

5.

X	-4	-3	-2	-1	0	1	2
$2x^2$	32	18	8	2	0	2	8
$4x-3$	-19	-15	-11	-7	-3	1	13
Y	13	3	-3	-5	-3	3	13

Plotting and linear scale

(b) X = 2. 6; x = 0.6

(c) Eq. of straight line = y = 3x + 3

6.   (a)   (i)

X	-3	-2.5	-2	-1.5	-1	-05	0	0.5	1	2	2.5
X3	-27	-15.63	-8	-3.38	-1	-0.13	0	0.13	1	8	15.63
X2	9	6.25	4	2.25	1	0.25	0	0.25	1	4	6.25
-2x	6	5	4	3	2	1	-	-1	-2	-4	-5
Y	-12	-4.38	0	1.87	2	1.12	0	-0.63	0	8	16.88

(ii) 0 < x < 1   -3 <x<-2

(b)   Line y = 2

(1.3, 1.3) and (-2, -2.3)

7.   a0   Find midpoint (centre) = $\frac{5 + (-1)}{2}$, $\frac{5 + (-3)}{2}$

=$[^4/_2 , ^2/_2]$

= (2, 1)

(b)   Vector of (a,b) = (2,1)

246

$$R = \begin{pmatrix} 5 \\ 5 \end{pmatrix} - \begin{pmatrix} 2 \\ 1 \end{pmatrix} = \begin{pmatrix} 3 \\ 4 \end{pmatrix}$$

$$\therefore r = \sqrt{3^2 + 4^2}$$

$$= 5 \text{ units}$$

$(x - 2)^2 + (y - 1)^2 = 5^2$

$X^2 - 4x + 4 + y^2 - 2y + 1 \quad = 25$

$X^2 + y^2 - 4x - 2y - 20 \quad = 0$

8. (a)   Let x be the number of computer bought. Using original price.

Original price per unit = $\dfrac{1800000}{X}$

New price per unit = $\dfrac{1800000 - 4000}{X}$

$\therefore \dfrac{1800000 - 4000}{X} = \dfrac{1800000}{x + 5}$

$1800000 x - 4000 x^2 + 9000000 - 20000x$

$= 1800000 x$

$X^2 + 5 x - 2250 = 0$

$X^2 + 50 x - 45x - 2250 = 0$

$X(x + 50) - 45 (x + 50) = 0$

247

$(x - 45)(x + 50) = 0$

$X = 45$

$\therefore$ No of computers bought = 50

(b)   No of computers left after breakage = 50- 2 = 49

Selling price to realize 15% profit

$= 1800000 \times 1.15 = 2070000$

Buying price per unit = $\underline{1800000}$

                         50

Profit per unit = 2070000   1800000

                      48   50

$= 43125 - 36\,000$

$= 7125$

9.   When x = 0, y = 2 $\therefore$ k x 1 x -2

$2 = -2k$

$K = -1$

CHAPTER EIGHTEEN: APPROXIMATION AND ERRORS

1. (a) $R = \dfrac{1}{0.000016} = \dfrac{1 \times 10^6}{1.6}$

   $= 625,000$

   (b) (i) Approximate value $= \dfrac{1}{0.00315 - 0.00313}$

   $= \dfrac{1}{0.00002} = \dfrac{1 \times 10^5}{2}$

   $= 50,000$

   (ii) Error $= 62500 - 50,000$

   $= 12500$

2. (a) $c = 2 \times 2.8 \times 22/7 = 17.6$

   $c/\pi = 17.6 \times 7/22 = 5.6$

   5.6 ± 0.05

   (b) $3.142 \times 2.8 \times 2 = 17.595$

   $3.142 \times 5.5 = 17.281$

   $3.142 \; 5.7 = 19.909$

   Limits $17.28 - 19.91$

. (a) Maximum possible Area

   $4.11 \times 2.21 = 9.083$

   Minimum possible Area

   $4.09 \times 2.19 = 8.9571$

   (b) Maximum possible wastage

9.0831 – 8.957

0.126m^2

4. (a) Working area = ½ x 6 x 4 =12

Maximum area = ½ x 6.5 x 4.5 = 14. 625

Minimum area = ½ x 5.5 x 3.5 = 9.625

Absolute error = 14.625 – 9.625

= 5

(b) % error = 5/12 x 100

= 41. 7%

Actual value = 788 x 0.006

5. 4.728

Approximate value = 800 x 0.006

= 4. 728

Approximate vale = 800 x 0.006

= 4.8

% Error   = 4.8 – 4.728 x 100

4.728

6. Greatest possible error = 64 (3. 15 – 3.05)

2

= $\dfrac{201.6 - 195 . 2}{2}$

= 3.2 cm^3

7. 40 ± 6.5

$\dfrac{6.5}{40} = 0.1625$

8. Min Perimeter = 74.75 cm

9. (i) Ans. 0.24 error 0.003

    (ii) Ans 0.23 error 0.007

10. Ans 10%

CHAPTER NINETEEN: TRIGONOMETRY I

1. $5/2 \; \theta = 210^0, 330^0$

   $\theta = \underline{420^0}, \underline{660}$

   $\quad 5 \quad\;\; 5$

   $= 84^0, 132^0$

2. (a) $X = 32 - 22$

   $\text{Tan } \theta = \underline{2}$

   $\phantom{\text{Tan } \theta = }\quad \sqrt{5}$

   (b) $\text{Sec}^2\, \theta = \tan^2 \theta + 1$

   $= 4/5 + 1$

   $= 1.8$

3. $\text{Sin}^2 (x - 30) \; t = \frac{1}{2} \times \frac{1}{2} = \frac{1}{4}$

   $\text{Sin} (x - 30) = \frac{1}{2} = \pm 0.5$

   $X = 30 = 30^0, 150^0, -30^0, -210^0$

   $X = 60^0, 180^0, 0^0, -120^0, -180^0$

4. $\text{Cos } 2x = \sin (90 - 2x)$

   $\text{Sin} (x + 30) = \text{Sin} (90 - 2x)$

   $S + 30 = 90 - 2x$

   $3x = 60$

   $X = 20^0$

   $\text{Cos}^2\, 3x = \text{Cos } 260$

   $= \left(\frac{1}{2} \right)^2$

   $= \frac{1}{4} \text{ or } 0.25$

5. $X^2 = (\sqrt{} = 12 = 4$

   $X = 2$

   (a) $\therefore \cos \alpha = \dfrac{2}{\sqrt{5}}$

   (b) $\tan(90 - \alpha) = 2$

6.   (a) $\sin^2 X + \cos X = 1$

      $\sin^2 x = 1 - \cos^2 x$

      $8(1 - \cos^2) + 2\cos X - 5 = 0$

      $8 - 8(\cos^2 x + 2\cos X - 5 = 0$

      $-8\cos^2 X + 2\cos X + 3 = 0$

      Let $\cos X$ be t

      $-8t2 + 2t + 3 = 0$

      Let $\cos x$ be t

      $-8t^2 + 2t + 3 = 0$

      $T = \tfrac{1}{2} \quad t = \tfrac{3}{4}$

      $\cos X = \tfrac{3}{4}$

   (b) $\tan X = \underline{\sqrt{7}}$

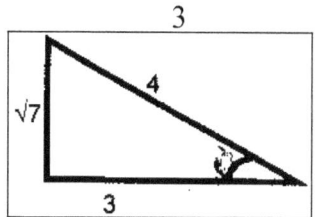

7.   Cos $2x^0$ = 0/8070

$2x^0$ = 36.2^0, 323.8^0, $396. 2^0$, 638.8^0

X^0 = 18.1^0, 161.9^0, 198.1^0, 341.9^0

8.   (a)   From $\triangle$ BCD

Sin 30^0 = $\underline{BD}$

12

BD = 12sin 30

= 12 x ½

= 6 cm

(b)   From $\triangle$ ABD

$\underline{Sin\ 45}$ = $\underline{sin\ \angle\ ADB}$

6   8

Sin $\angle$ ADB = $\underline{8\ sin\ 45}$

6

= $\underline{4\ x\ 0.7071}$

3

= 0.9428

$\angle$ADB = 70.53

9.

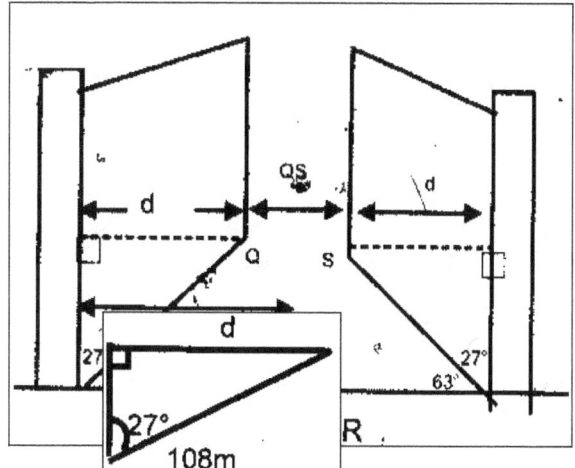

Where 2d + Qs = 3.6m

Qs   =   3.6 – 2d

$\underline{1.8}$ = $\underline{d}$

$\mathrm{Sin}\ 90^0$ $\mathrm{Sin}\ 27^0$

D = 1.8 sin 27^0

        $\mathrm{Sin}\ 90^0$

= 0.8172

QS = 3.6 – 1.6344

= 1.9656m

10. $\mathrm{Tan}\ 30^0 =$ AE

                100

(a)   (i)   AE = 100 Tan 30^0 = 57.74m

       (ii)   57.74 = AC = 81.6m

            $\mathrm{Sin}\ 45^0$ $\mathrm{Sin}\ 90^0$

$$AD^2 = 80^2 + 81.66^2 - 2(80 + 81.66) \cos 100$$

$$= 6400 + 6668.36 - 2(161.66) \cos 100$$

$$= 13124.48$$

$$AD = 114.6m$$

(iii) $\cos 30^0 = \dfrac{100}{AB}$; $AB = \dfrac{100}{\cos 30^0} = 115.47$

$$EC = 57.74m \; (\angle AEC \text{ is isosceles})$$

$$\text{Perimeter} = BE + EC + CD + DA + AB$$

$$= 100 + 57.74 + 80 + 114.6 + 115.5$$

$$= 487.84$$

(b) $487.84 - 2.8 = 485.04$

$$\dfrac{485.04 \times 5}{480} = 5.0525$$

∴   6 rolls of barbed wire are required

11    $l^2 = 5^2 - (2\sqrt5)^2 = 5$

$L = \sqrt5$

∴ $\tan(90 - x)^0 = \dfrac{2\sqrt5}{\sqrt5}$ or 2

12. $\dfrac{1}{2} \angle ACB = 38.5^0$

$$\dfrac{8.4}{\sin 100.5} = \dfrac{x}{\sin 41^0}$$

$X = \dfrac{8.4 \sin 41^0}{}$

Sin 100.5

X = CN = 5.6

13.   (a) $\angle ABQ = 180 - 955^0 = 845^0$

        ∴ AB = $\dfrac{5.8}{\text{Cos } 84.5}$ = 60.5 m = 14.5^0 = 61m

    (b)   (i) $\angle ABC = 95.5 + (90.30.5)$

            $= 155^0$

            Scale 1cm: 10cm

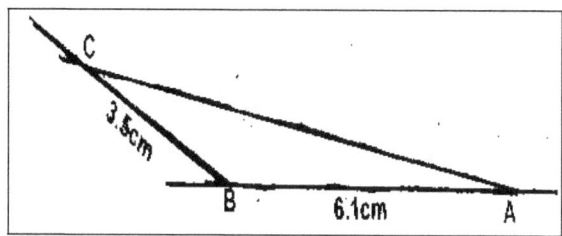

            AC = 9.4 x 10 = 94m

            (Using 63 m = 96m) ± 1m

        (ii) $\angle BCA = 16^0 \pm 1^0$

            ∴∠ of depression of A from C

            $= 30.5^0 - 16^0$

CHAPTER TWENTY: SURDS AND FURTHER LOGARITHMS

1. $\log x^3 + \log 5x = 5 \log 2/5$

    $\log (x^3 \times 5x) = \log \dfrac{32 \times 5}{2}$

    $5x^4 = 80$

    $X^4 = 16$

    $X = 2$

2. $(1 + \sqrt{3})(1 - \sqrt{3}) = 1 - 3 = -2$

    $\dfrac{1}{1 + \sqrt{3}} \quad x \quad \dfrac{1 - \sqrt{3}}{1 - \sqrt{3}}$

    $= \dfrac{1 - \sqrt{3}}{-2} = -\frac{1}{2} + \dfrac{\sqrt{3}}{3}$

    $\dfrac{1.7321 - 0.5}{2}$

    $= 0.366$

3. $\dfrac{\sqrt{14}\sqrt{7} + \sqrt{2}) - \sqrt{14}\sqrt{7} - \sqrt{2})}{(\sqrt{7} - \sqrt{2})(\sqrt{7} + \sqrt{2})}$

    $a = 4/5, b = 0$

4. $49^{(x+1)} + 7^{(2x)} = 350$

    $49(7^{2x}) + 7^{(2x)} = 350$

    $50(7^{(2x)}) = 350$

$7^{(2x)} = 7$

$2x = 1$

$X = \frac{1}{2}$

5. $5 \log \underline{1}\ x^2 = \log \underline{1}$

    $\ 125 \ 125$

    $1\ x^2 \quad = 1$

    $\overline{125} \quad \overline{125}$

    $X^2 = 1$

    $X = 1$

6. $\dfrac{\sqrt{14} + 2\sqrt{3} - (\sqrt{14} - 2\sqrt{3})}{(\sqrt{14})^2 - (2\sqrt{3})^2} = \dfrac{4\sqrt{3}}{2}$

7.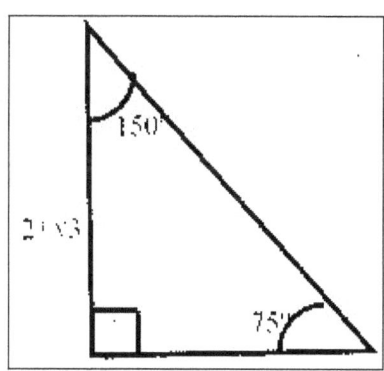

Tan $15^0 = \dfrac{1}{2 + \sqrt{3}}$

$\dfrac{1}{2 + \sqrt{3}} \quad x \quad \dfrac{2 - \sqrt{3}}{2 + \sqrt{3}}$

$= 2 - \sqrt{3}$

8. $\dfrac{3\sqrt{7} + 6\sqrt{2}}{4\sqrt{2} + 2\sqrt{7}}$

$\dfrac{(3\sqrt{} + 6\sqrt{2})(4\sqrt{2} - 2\sqrt{7})}{(4\sqrt{2} + 2\sqrt{7}(4\sqrt{2} - 2\sqrt{7})}$

$\dfrac{12\sqrt{14} - 42 + 48 - 12\sqrt{14}}{32 - 38}$

$= {}^6/_4$ or 1.5

9. $\dfrac{3}{\sqrt{5} - 2} + \dfrac{1}{\sqrt{5}} \quad = \quad \dfrac{3(\sqrt{5} + 2)}{5 - 4} + \dfrac{1}{5}\sqrt{5}$

$= 3\sqrt{5} + 6 + \dfrac{1}{5}\sqrt{5}$

$= \dfrac{6 + 16}{5}\sqrt{5}$

10. $y = 0$

11. $x = 5, x = 1/3$

12. $x = 1, x = 0$

13. $x = 3$

CHAPTER TWENTY ONE: COMMERCIAL ARITHMETIC II

1. (a) by 30th June, 1996

    A = 12000 x 0.9

    = Kshs 13080

    (b) By 30th June 1997

    A = 12000 x 1.09^2

    13080 + 14257.20

    Kshs 27337.20

2. (a) Cost/ ton/km = $\dfrac{24000}{28 \times 48}$

    Kimani received

    $\dfrac{24000 \times 96 \times 49}{28 \times 48}$

    = 84000

    (b) Profit = $84000 - \dfrac{96}{8} \times 3000 = 48,000$

    (c) Achieng received $\dfrac{84}{28} \times 24,000 = 72,000$

    Transportation cost = $72,000 \times \dfrac{100}{144} = 50,000$

3. (a) Total earning $\dfrac{40480}{20}$

    435 x 2 =   870

$435 \times 3 = \quad 1305$

$435 \times 4 = \quad 1740$

$435 \times 5 = \quad 2175$

$284 \times 6 = \quad \underline{1704}$

$\qquad\qquad\qquad 7794$

(b) Net tax – Kshs 7794 – 800

Kshs 6994

(c) New earnings

$1.5 \times 2024 = £3036$

$£ 3036 - £ 2024 = £ 1012$

Net tax $= 1012 \times 6$

$= $ Kshs 6072

% age excess $= 60\ 72 \times 100$

$\qquad\qquad\qquad 7794$

4. (a) (i) $750,000 \times \dfrac{90}{100}$

$= 675,000$

   (ii) $675,000\,(1.1)^3 = 898,425$

$898,425 + 75,000 = 973,425$

(b) $675,000\,(1.1)_n = 816,750$

$(1.1)^n = 1.21$

$N = \quad \underline{0.0828}$

0.0414

N = 2 years

5. $S1 = \dfrac{P \times 2 \times 5}{100}$

    $= 0.1P$

    Amount after 2 years $= \dfrac{P(1+5)^2}{100}$

    $P(1.05)^2 = 1.1025P$

    Compound interest $= 1.1025P - p$

    $= 0.1025P$

    $0.1025P - 0.1P = 0.0025P = 210$

    $\dfrac{P\ 210 \times 10^4}{0.0025 \times 10^4} =$ 84 000

6. (a)   (i) $A = P + 1$

                Total interest $= 12,800 \times 3$

                              $= 38,400$

                $P = A - 1$

                $P = 358,400 - 38,400$

                $= Kshs\ 320,000$

        (ii) $\dfrac{R}{100} = \dfrac{1}{PT}$

              $R = \dfrac{1 \times 100}{PT}$

$$= \frac{12,800 \times 100}{320,000}$$

$$R = 4\%$$

(b)   Deposit $= \frac{25}{100} \times 56\ 000 = 14,000$

Balance $= 56,000 - 14,00 = 42,000$

$$\frac{42,000}{N} = 2625 \qquad n = 16 \text{ installments}$$

(ii)   B.P $= \frac{175}{200} \times 40,000 = $ Kshs 35,000

Difference $= 56,000 - 35,000$

$$= 21,000$$

$$\frac{21,000}{35,000} \times 100 = 60\%$$

7.   Let monthly income be y

Taxable income	Rate	Tax payable	Acc. Tax
9681	10%	10% x 9681	968.10
18,801 – 9681 = 9120	15%	15% x 9120 = 1368	2336.10
$\therefore y - 9684 = x$	15%	15% x = 947. 90	1961

$X = \frac{94.90}{15} \times 100$

$= 6319. 3$

$Y = 9681 = 6319.3$

$Y = 6319.3 + 9681$

= 16000.30

= Kshs 16,000

8.   Interest = (13 800 – 2280) x $\underline{20}$ x 2

                                100

= 11520 x 0.2 x 2 = 4608

Each monthly installments = $\dfrac{11520 + 4608}{24}$

Kshs 672

10.   (a) Kshs 60, 000

        (b) Kshs 79, 860

11.   Kshs 240, 000

12.   Amount payable = Kshs 75510

13.   = 200,00

14.   9663.6

CHAPTER TWENTY TWO: CIRCLES, CHORDS AND TANGENTS

1. Area of the sect or = $\dfrac{75}{360}$ x $\dfrac{22}{7}$ x 14 x 14

    = 128. 3 cm^2

    Area of Δ = ½ x 14 x 14 sin 75^0

    = ½ x 14 x 14 x 0.9659

    = 94.64 cm^2

2. (a)   PS = $(34^2 – 16^2)$ = 900

        = 30

    (b)   Cos POS = $\dfrac{17^2 + 17^2 – 30^2}{2 \text{ x } 17 \text{ x } 17}$ = $\dfrac{-322}{578}$

        = -0.5572

    ∴ POS = 123^0 50 (123. 86^0)

3. (a)   6 x C = 4.8 x 5

        XC = $\dfrac{4.8 \text{ x } 5}{6}$ = 4

    (b)   BT^2 = (6 + 4 + 8) x 8

        = 18 x 8 = 144

        BT = 12

4. (a)   Area = $\dfrac{120}{360}$ x 7 x 7 x $\dfrac{22}{7}$ = 51 1/3 cm^2

    (b)   ½ AD = 7sin 600 = 7 cos 60

AB = 14 -2 x 7x 0.5 = 7

Area of trapezium XZBY = ½) (7 + 14) x 6.062

= 63.65 cm²

(c)   Area of shaded region = 2 (63. 65 – 511 1/3)

= 127. 30 – 102. 67

= 24. 63 cm²

5.   θ = Angle POT

Cos θ = 7/25

θ = 73⁰ 55' or 73. 74

$\theta = 73^0\,55'$ or 73. 74

PQ = 7 x 2 sin 73. 74

= 14 x 0.9608

= 13. 44 cm

6.

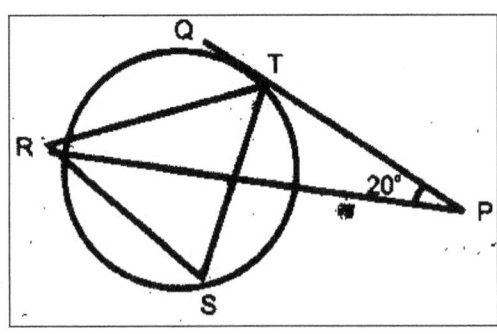

< RST = 35 + 20 = 55

= 55⁰

7.   Area of each sector

$\underline{60}$ 360 x π x 6²

360

= 18.849555592

Area of Δ = ½ x 6 x 6 x sin 60⁰

= 15. 5884527

∴ Area of the shaded region

15. 588445727 + 2(18.84955592) − 15.5884527)

= 15. 58845727 + 6.522197303

= 22.11065457

= 22.11

8.   (a) $NR = 4^2 + 7.5^2$

        = 8.5 cm

    (b) $QR (14 + 8.5) = 7.5^2$

        $4xAN = 14 (8.5 − 2.5)$

        $AN = \dfrac{14 \times 6}{4}$

        = 21 cm

CHAPTER TWENTY THREE: MATRICES

1. $A^2 = $ 1 2   1 2 $=$ 9 8

   4 3   4 3   16 17

$$B = \begin{pmatrix} 9 & 8 \\ 16 & 17 \end{pmatrix}\begin{pmatrix} 1 & 2 \\ 4 & 3 \end{pmatrix} = \begin{pmatrix} 8 & 6 \\ 12 & 14 \end{pmatrix}$$

2. $\begin{pmatrix} 1 & 3 \\ 5 & 3 \end{pmatrix}\begin{pmatrix} 3 & 1 \\ 5 & -1 \end{pmatrix} - \begin{pmatrix} 3 & 1 \\ 5 & -1 \end{pmatrix} = \begin{pmatrix} p & 0 \\ 0 & q \end{pmatrix}$

   $18 = 3p \quad 5q = 30$

   $P = 6 \quad q = 6$

3. (a) x^2 0 -   x 0 $=$ x^2 0

      5   y   5 y $5x + 5y$ y^2

   (b) $\begin{pmatrix} x^2 & 0 \\ 5x + 5y & y^2 \end{pmatrix} = \begin{pmatrix} 1 & 0 \\ 0 & 1 \end{pmatrix}$

   $5x + 5y = 0$

   If $x = \begin{pmatrix} 1, & y \end{pmatrix} = -1$

If x = -1, y = 1

4. (a) m = 54 – 56 = -2

Inverse matrix = -1 $\begin{pmatrix} 6 & -8 \\ -7 & 9 \end{pmatrix}$

$\begin{pmatrix} & 2 & \\ & & \end{pmatrix}$

Or $\begin{matrix} -3 & 4 \\ \dfrac{7}{2} & -\dfrac{9}{2} \end{matrix}$

(b) Let the price of each bicycle be x and each radio be y

36 x + 32y = 227280

28x + 24 y = 174960

$\begin{pmatrix} 36 & 32 \\ 28 & 24 \end{pmatrix} \begin{pmatrix} x \\ y \end{pmatrix} = \begin{pmatrix} 227280 \\ 174960 \end{pmatrix}$

$\begin{pmatrix} 9 & 8 \\ 7 & 6 \end{pmatrix} \begin{pmatrix} x \\ y \end{pmatrix} = \begin{pmatrix} 56 & 820 \\ 43 & 740 \end{pmatrix}$

$\begin{pmatrix} 6 & -8 \\ -7 & 9 \end{pmatrix} \begin{pmatrix} 9 & 8 \\ 7 & 6 \end{pmatrix} \begin{pmatrix} x \\ y \end{pmatrix} = \begin{pmatrix} 6 & -8 \\ -7 & 9 \end{pmatrix} \begin{pmatrix} 56 & 820 \\ 43 & 740 \end{pmatrix}$

271

$$\begin{pmatrix} -2 & 0 \\ 0 & -2 \end{pmatrix} \begin{pmatrix} x \\ y \end{pmatrix} = \begin{pmatrix} -9000 \\ -4080 \end{pmatrix} \begin{pmatrix} 1 & 0 \\ 0 & 1 \end{pmatrix} \begin{pmatrix} x \\ y \end{pmatrix}$$

$$\begin{pmatrix} 4500 \\ 2040 \end{pmatrix}$$

$\therefore$ each bicycle price = 4500 x 0.9 = 4050

New price of radio = 2040 x 1.1 = 2244

$$\therefore (64 \quad 56) \begin{pmatrix} 4050 \\ 2244 \end{pmatrix} = (259200 + 125664)$$

$\therefore$ Total cost in 3rd week

= 259200 + 125664 = 384864

5. $T-1 = \begin{matrix} 1/3 & 2/3 \\ 1/3 & -1/3 \end{matrix}$

Coordinates (3,2)

6. (i) $\begin{pmatrix} 1 & -1/2 \\ -1/2 & 5 \end{pmatrix}$

    (ii) $\begin{pmatrix} -1/3 & -2/3 \end{pmatrix}$

272

$$\begin{array}{cc} 4 & -8/3 \end{array}$$

(iii) $\begin{pmatrix} -3/2 & -1 \\ 0 & 7/2 \end{pmatrix}$

7. $k = 3$

8.   Shirts cost Kshs 120

Trousers cost Kshs 240

CHAPTER TWENTY FOUR: FORMULAE AND VARIATIONS

1. $$\frac{1}{Sc^2} = \frac{3V + 2}{2\pi^3}$$

$$\frac{C^2 = 2\pi r^3}{3SV + 4\pi r^3 S}$$

$$\sqrt{\frac{C = 2\pi r^3}{3SV + 4\pi r^3 S}}$$

2. $$\frac{2T}{M} = V^2 - r^2$$

$$V^2 = v^2 = v^2 - \frac{2T}{M}$$

$$V = V^2 - \frac{2T}{M}$$

3. $y(Cx^2 - a) = b - bx^2$

$X^2(yc + b) = b + ya$

$$X^2 = \frac{b + ya}{Yc + b}$$

$$X = \frac{b + ya}{Ya + b}$$

4. $\log y = \log(10x^n)$

$\text{Log } y = \log 10 + n \log x$

$n \log x = \log y - \log 10$

$$n = \frac{\log y - \log 10}{\log x}$$

5. (a) $T = a + b\sqrt{s}$ or $T = b + a\sqrt{s}$

(b) $a + b\sqrt{16} = 24$

$a + b\sqrt{36} = 32$

$a + 3b = 24$

$\underline{a + 6b = 32}$

$-2b = -8$

$b = 4 \qquad a = 8$

6. $P = \underline{k} + c$

$\qquad q$

$10 = \underline{k} + C = \underline{k} + 1.5c$

$\qquad 1.5 \qquad\quad 1.5$

$K + 1.5c = 15$

$20 = \quad\underline{k}\quad + \quad c = \underline{k} + 1.25$

$\qquad 1.25 \qquad\qquad 1.25$

$K = 1.25c = 25$

$K + 1.5c = 15$

$\underline{K + 1.25c = 25}$

$0.25c = -10$

$C = -40,\ k = 75$

7. $px - py = xy$

$Px = xy + py$

$Px = xy + py$

$Px = y(x+p)$

$Y = \dfrac{px}{X + p}$

8. $p^2 = p^2 - pr - pq + qr$

$Pr + pq = qr$

$P(r+q) = qr$

$P = \dfrac{qr}{r + q}$

9. $D = \dfrac{km}{R^3}$

$2 = \dfrac{k}{125} \times 500 \Rightarrow k = \frac{1}{2}$

$D = \dfrac{1}{R^3} \times \dfrac{m}{2} = \dfrac{m}{2r^3}$

$R^3 = \dfrac{540}{2 \times 10} = 27$

$R = 3cm$

10. $\sqrt{p} = r\,1 - as^2$

$P = r^2(1 - as^2)$

$\underline{P} = 1 - as^2)$

276

$$R^2$$

$$as^2 = 1 - \frac{P}{r^2} \quad = \frac{r^2 - p}{r^2}$$

$$S^2 \quad = \quad \frac{r^2 - p}{a \; r^2}$$

$$S \quad = \sqrt{\frac{r^2 - p}{a \; r^2}}$$

11. $\quad t \quad a \quad x$

$\quad\quad\quad\quad Y$

$$\therefore t = \sqrt{\frac{kx}{y}}$$

$$t_1 = \sqrt{\frac{kx_1}{y_1}}$$

$$t_2 = \sqrt{\frac{k.\,096x_1}{1.44y_1}} \quad = \sqrt{\frac{k0.96x_1}{1.2\,y_1}} \quad = \sqrt{\frac{0.8kx_1}{y_1}} \quad = 0.8t_1$$

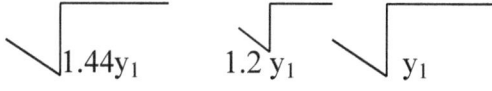

$$\% \text{ Decrease} \quad = \frac{t_1 - t_2}{t_1} \times 100$$

$$= \frac{t_1 - 0.8t_1}{} \times 100$$

$$\frac{t_1}{} = 20\%$$

12.   (a)   (i) $y = \dfrac{k}{x^n}$

            (ii) $K = 12 \times 2$ and $K = 3 \times 4^n$

                 $12 \times n^2$ and $k = 3 \times 4^n$ or $\dfrac{k}{3} = \dfrac{k^2}{144}$

                 $2^{n+2} = 2^{2n}$ or $k^2 - 48k = 0$

                 $N + 2 = 2n$ or $k(k - 48) = 0$

                 $N = 2$ or $k = 48$

                 $K = 48$ or $K = 48$

                 $K = 48$ or $n = 2$

   (b) $y = \dfrac{48}{\left(5\frac{1}{3}\right)^2} = 1\frac{11}{16}$ or 1.6875

                 $= 1.688$

13.   9/8 Ohms

14.   (a) $V = 52.5r^2 + 2.1r^3$

         (b) $974.4\ cm^3$

         (c)   25

15.   9.6 kg

16. $X = \dfrac{p^2z}{Y - p^2}$

17. (a) $C = a + \dfrac{b}{N}$ where a + b are consonants

    (b) Fixed charge, a = Kshs 8000

    (c) 70 people

18. A = 79,(-78.82)

19. $P = \dfrac{A^2 N}{E^2 - n^2}$

CHAPTER TWENTY FIVE: SEQUENCE AND SERIES

1. $10, 10 + 2d, 10 + 6d$

    $$\frac{10d + 2d}{10} = \frac{10d + 6d}{10 + 2d}$$

    $100 + 40d + 4d^2 = 100 + 60d$

    $4d^2 - 20d = 0$

    $D = 5$ or $d = 0$

2. (a) 2nd year saving = $2000 \times \dfrac{115}{100}$

    = Kshs 2300

    (b) 3rd year saving = $2300 \times \dfrac{115}{100}$

    = Kshs 645

    (c) Common ration = $\dfrac{115}{100}$ or $\dfrac{23}{10}$

    (d) $\dfrac{2000\,(1.15 - 1)}{1.15 - 1} = 58000$

    $2000 \times 1.15^{n} = 8700 + 2000$

    $1.15^{n} = \dfrac{(8700 + 2000)}{2000}$

    $n \log 1.15^{n} = \log 5.35$

$0.0607^n = 0.7284$

$N = \dfrac{0.7284}{0.0607} = 11.99$

$= 12$

(e)　$S_{20} = \dfrac{2000\,(1.15^{20} - 1)}{1.15 - 1}$

$= \dfrac{2000 \times 16.37 - 2000}{0.15} = \dfrac{30.730}{0.15}$

$= 204800$

$= 204933$

3. 　(a)　$\dfrac{ar^2}{a + ar} = \dfrac{16}{12} = \dfrac{4}{3}$

　　　　Ratio = 4:3

　　(b)　$3r^2 - 4r - 4 = 0$

　　　　$3r^2 - 6r - 2r - 4 = 0$

　　　　$(3r + 2)\,(r - 2) = 0$

　　　　$R = 2/3$ 0r $r = 2$

　　　　$\therefore r = {}^{-2}/_3$

4. 　(a)　$^n/_2\,(4 + 20) = 252$

$N = {}^{504}/24 = 21$

$^{21}/_7\,(2 \times 4 + (21 - 1)\,d = 252$

$21\,(8 + 20d) = 504$

$D = {}^{16}/_{20} = {}^4/_5$

(b) $50 \times 1.8^n = 1200000$

$N \log 1.8 = \log \dfrac{1200000}{50}$

$N \times 0.2553 = 4.3802$

$= 4.3802$

$= 0.2553$

$= 17.16$

Time taken 17.16×20

$= 343.2$ minutes $(5.72\ h)$

5. (a) $T_{40} = 500 + (40-1)\ 50$

$= 500 + 1950$

$= 2450$

(b) $S40 = 40/2\ (500 \times 2 + (40 - 1)\ 50$

$= 20\ (1000 + 1950$

$= 59{,}000$

6. $\underline{67 - 32}$

14

$= 2.5$

$= 67.6 \times 2.5$

$= 52\ cm$

7. (a) = 32, r = ½

8. (a) d = 5; a = 10

    (b) p > 119/5

9. (a) 5, 7, 9, 11

    (b) 2700

    (c) n = 24

CHAPTER TWENTY SIX: VECTORS

1.   (a)   (i) AV = AD + DV = a + c

                (ii) BV = BA + AV = a + c – b

    (b)   BO = ½ BD = ½ (a – b)

            OV = OB + BV

                = ½ (b – a) + a + c –b

                = ½ a + c – ½ b

            OM = $^3/_7$ OV

= 3/7 (½ a + c – ½ b)

BM = BO + OM

½ (a-b) + 3/7 (½ a + c – ½ b)

$$= \frac{7a - 7b + 3a + 6c - 3b}{14}$$

$$\frac{10a - 10b + 6c}{14}$$

= 1/7 (5a – 5b + 3c)

2.   (a)   (i)   AB = b – a

            (ii)   AP = 5/8 (b- a)

            (iii)   BP = 5/8 (a- b)

            (iv)   OP = OA + AP or OB + BP

                    = a + 5/8 (b –a)

                    = 5/8 a + 5/8b

    (b)   OP = 5/8 + 5/8b

$OQ = a - 5/8\ a + 9/40b$

$= 3/8a + 9/40b$

$OQ: OP = 3/8a + 9/40b: 5/8a + 3/8b$

$= 3/8(a+ 3/5b) : 5/8(a+ 3/5b)$

$OQ: QP = 3:2$

3. (a) (i) $AN = OM - OA$

$^4/_5b - a$

(ii) $BM = OM - OB$

$^2/_5a - b$

(b) (i) $AX = sAN$

$= s(^4/_5\ b - a)$

$= ^4/_5\ sb - sa$

$BX - tBM$

$= t(^2/_5a - b)$

$= ^2/_5ta - tb$

(ii) $OX = OA + AX$

$= a + 4/5\ b5 - as$

$= a\ (1-5) + 4/5\ sb$

$OX = OB + BX$

$B + 2/5at - bt$

$= 2/5\ ta - b\ (l-t)$

$\therefore a\ (1 - s) + 4/5sb = 2/5ta - b\ (l-t)$

$\therefore l- S = 2/5t$

285

And

$4/5 \, S = 1 - t$ ....(ii)

From equal (ii)

$S = (1 - t) \, 5/4$

$= 5/4 - 5/4 \, t$

Substituting in l

$L - S = 2/5 \, t; \; l = 2/5 \, t + S$

$L = 2/5t + 5/4 - 5/4 \, t$

$5/4 \, t - 2/5t = 5/4 - 1$

$\underline{17t} = \underline{1}$

$20 \qquad 4$

$T = {}^{5}/_{17}$

$S = {}^{10}/_{17}$

4. $PQ = \quad 3i \qquad 4i \qquad -1$

$6j - 3j = 9j$

$6k \;\; 2k \;\; 4k$

$PQ = \quad (-1)^2 \; + (9)^2 + (4)^2$

$= \sqrt{98}$

$= 7\sqrt{2}$

$|PQ| = \sqrt{1^2 + (9)^2 + 4^2}$

$= 7\sqrt{2}$

5. (a) $OR = r - \frac{3}{2}p$

   $PS = 2r - p$

   (b) $OK = \frac{2}{3}p + m(r - \frac{3}{2}p)$

   $OK = p + n(2r - p)$

   $\frac{3}{2}p + m(r - \frac{3}{2}p) + n(2r - p)$

   $2n = m \ \ldots\ldots(i)$

   $\frac{3}{2}, -\frac{3}{2} = 1 - n \ \ldots.. (ii)$

   $M = \frac{1}{2} \ \ n = \frac{1}{4}$

   (c) $PK : KS = 1:3$

6. $OA = \begin{pmatrix} 1 \\ -1 \\ 1 \end{pmatrix}$

$OT = 2 \begin{pmatrix} \\ 0 \end{pmatrix}$

1.5

Let $OB = \begin{pmatrix} x \\ Y \\ Z \end{pmatrix}$

$$\underline{X + 1 = 2}; \qquad \underline{y + (-1) = 0}; \underline{Z + 1} = 1.5$$
$$2 \qquad\qquad 2 \qquad\qquad 2$$

$$X + 1 = 4' \; y - 1 = 0; \; z + 1 = 3$$

$$X = 3; \; y = 1; \; z = 1$$

$$\therefore OB = \begin{pmatrix} 3 \\ 1 \\ 2 \end{pmatrix}$$

OB = 3i + j + 2k

7.

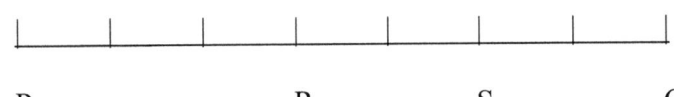

P   R   S   Q

PR: RQ = 3: 4

PS : SR = 5: -2

PQ = 8 cm

RS = 2/7 PQ

= 2/7 x 8

= 2.29cm

8.   (a)   OT = 12/7p + 3/7r

                 QT = 3/7r – 9/7p

                 = 3/7 (r-3p)

(b)   (i)   QR = r – 3p

QT = 3/7QR

∴ QT & QR are parallel and Q is a common point

∴ Q, T and R lie on a straight line

(ii) QT : TR = 3:4

∴ T divides QR in the ratio 3:4

9. $8 – k = -3$

$K – 3$

$8 – k = -3 k + 9$

$2k = 1$

$K = ½$

Taking a general point (x, y)

$Y – 8 = -3$

$X – ½$

$Y – 8 = -3x + 3/2$

$3x + y = 9 ½$ or $6x + 2y = 19$

10. $q^2 + (1/3)^2 + (2/3)^2 \quad = 1$

$q^2 + 1/9 + 4/9 \qquad = 1$

$q^2 + 5/9 \qquad\qquad = 1$

$q^2 = 4/9$

∴ q = 2/3

11.   (a) OL = 3OA

= 3 (1, 6)

= 3, 18

ON = 2/3 OB

= 2/3 (15, 6)

= (10, 4)

$\therefore LN = 10 \begin{pmatrix} -3 \\ 4 \end{pmatrix} \begin{pmatrix} = \\ 18 \end{pmatrix} \underline{7} \\ 14$

(b) LM = 3/7 LN = 3/7 (7) = (3)

(-14)   (-6)

Let co- ordinates of M be (x, y)

$$\begin{pmatrix} x \\ y \end{pmatrix} - \begin{pmatrix} 3 \\ 18 \end{pmatrix} = \begin{pmatrix} 3 \\ -6 \end{pmatrix}$$

x – 3 = 3 $\therefore$ x = 6

y – 18 =-6 $\therefore$ y = 12

Hence M (6 , 12)

(c) (l) $\underline{6}$ OT = OM

$\quad 7$

$$\underline{6} \begin{pmatrix} x \\ y \end{pmatrix} = \begin{pmatrix} 6 \\ 12 \end{pmatrix}$$
$\quad 7$

$\underline{6x} = 6$ $\therefore$ x = 7

7

$\underline{6y} = 12$ $\therefore$ y = 14

 7

$$\therefore OT = \underline{7}$$
$$14$$

(ii) $LT = \begin{pmatrix} 7 \\ 14 \end{pmatrix} - \begin{pmatrix} 3 \\ 18 \end{pmatrix} = \begin{pmatrix} 4 \\ -4 \end{pmatrix}$

$BT = \begin{pmatrix} 15 \\ 6 \end{pmatrix} - \begin{pmatrix} 7 \\ 14 \end{pmatrix} = \begin{pmatrix} 8 \\ 8 \end{pmatrix}$

BT = 2 LT and they share point T

2007

12.   (a)   (i) XR = r – $\underline{1}$q
                        3

(ii) YQ = q – $\underline{3}$ r
                        7

      (b)   (i) OE = $\underline{1}$q - $\underline{1}$ mq + mr
                        3   3

             (ii) OE = $\underline{3}$ r – $\underline{3}$ nr + nq
                        7   7

      (c)   OE = $\underline{1}$ q + m (r – $\underline{1}$)q
                    3   3

             = $\underline{3}$ r + n (q – $\underline{3}$r)
                 7   7

291

$$\begin{pmatrix} \dfrac{1}{3} - \dfrac{1}{3}\,m \\ 3 \quad 3 \end{pmatrix} \quad q + mr = nq + \left(\dfrac{3}{7} - \dfrac{3}{7}\,n\right) r$$

$$\dfrac{1}{3} - \dfrac{1}{3}m = n$$

$$M = \dfrac{3}{7} - \dfrac{3}{7}\,n$$

$$M = \dfrac{3}{7} - \dfrac{3}{7}\left(\dfrac{1}{3} - \dfrac{1}{3}\right) m$$

$$M = \dfrac{3}{7} - \dfrac{1}{7} + \dfrac{1}{7}\,m \qquad m = \dfrac{1}{3}$$

$$N = \dfrac{1}{3} - \dfrac{1}{3} \times \dfrac{1}{3} = \dfrac{2}{9}$$

13. $|P| = \sqrt{3^2 + (-1)^2 + \left(1\,\dfrac{1}{2}\right)^2} = 3.5$

$Q = 2p$

$Q = 6i - 2j + 3k$ or $6i + 2j - 3k$

14. $19i - 5j$

15. $KL - 3NM = 3u$

    $KL = KN - NM$

    $3i = w + u + v$

    $2u = w + v$

16. (a) $4j - j + 7k$

    (b) $\sqrt{66} = 8.124$

17. $(-9.5, -4)$

18. (a) $b - a - \frac{2}{3}b$

    (b) (i) $k(a - \frac{2}{3}b)$

    (ii) $k = 2, m = 1$

19. (a) (i) $AC = a + b$

    (ii) $AC = a - \frac{2}{3}b$

    (b) $\frac{2}{3}a - \frac{8}{9}b = \frac{2}{3}(a - \frac{4}{3}b)$

    (c) $k = 8, h = 22$

    $PX : RX = 1:7$

20. $I + j + k$

21. $P = 19.7$

CHAPTER TWENTY SEVEN: BINOMIAL EXPRESSION

1. $146 \times 15 + 15x + 20x + 6x + x$

    $1 + 6(0.03) + 15 (0.03) + 20(0.03)$

    $= 1 + 0.18 + 0.135 + 0.0054$

    $= 1.19404$

    $= 1.194$

2. $10(0.96) = (1-0.04)$

    $= 1 + 5 (-0.04) + 10 (-0.04) + 10(-0.04)$

    $= 1 - 0.2 + 0.016 - 0.00064 + 0.0000128 + 0.000001024$

    $= 0.81536$

    (0.8153728 or 8153726976)

3. $(3x - y) 4 \Rightarrow (3x^4 y^0, (3x)^3 y, (3x)^3 y, (3x)^2 y^2$

    $(3x)y^3, (3x)^0 y^4$

    $(3x-y)^4 = 81x^4 - 108x^3 y + 54x^2 y^2 - 36xy^3 + y^4$

    $X = 2$ and $y = 0.2$

    $(6 - 0.2)^4 = 81(2)^4 - 108(2)^3 \times 0.2 + 54(2)^2 \times 0.22$

    $162 - 43.2 + 86.4$

    $= 205.2$

4. (a) C.d $= 64800 - 60000 = 69600 - 64800 = 4800$

    $A = 60000$

    N^{th} term $= a + (n-1)d$

    $= 60000 + (n-1) 4800$

(b)   Common ration = $\dfrac{64800}{60000}$ = $\dfrac{69984}{64800}$ = 1.08

Nth term = ar(n-1) where a = 60000

R = 1.08

= $60{,}000\,(1.08)^{(n-1)}$

(c)   7th term

Andi = 60000 + (7-1) 4800

= 88800

Amoit = $ar^{(n-1)}$ = $60000\,(1.08)^6$

= 95213

Difference = 95213 – 888000

Kshs 64.13

5.   Let $\dfrac{1}{\sqrt2}$ be a

$(2 + a)\,5 + (2 -a)\,5$

$(2 + a)^5 = 2^5 + 5\,(2^4 a) + 10\,(2^3 a^2) + 10(\,2^2 a^3) + 5\,(2a^4) + a^5$

$= 32 + 80a + 80a^2 + 40^3 + 10a^4 + a^5$

$(2 + \dfrac{1}{\sqrt2})^5$ $= 32 + \dfrac{80}{\sqrt2} + 40 + \dfrac{20}{\sqrt2} + \dfrac{5}{\sqrt2} + \dfrac{1}{4\sqrt2}$

$(2 - a)^5 = 32 - 80a + 80a^2 - 40a^3 + 10a^4 - a^5$

$(2 - \underline{1})^5 = 32 - \underline{80} - 40 - \underline{20} + \underline{5} - \underline{1}$

$\quad\quad \sqrt{2} \quad\quad\quad \sqrt{2} \quad\quad \sqrt{2} \quad \sqrt{2} \quad 4\sqrt{2}$

$\left(2 + \dfrac{1}{\sqrt{2}} \right)^5 + \left(2 - \dfrac{1}{\sqrt{2}} \right)^5 = \left. \right) \; 32 + \; 32 \; + 40 \; + 40 \; + 5/2 + 5/2$

$= 149$

6.   (a) $1.1^5 \; \underline{1}x + 5.1^4 \; \underline{1}x + 10.1^3 \quad \underline{1}x \quad + 10.1^2$

$\quad\quad\quad\quad 2 \quad\quad\quad\quad 2 \quad\quad\quad\quad 2$

$\left(1x \atop 2 \right)^2 + 1.1^0 \left(1x \atop 2 \right)^5$

$1 + 5/2\,x + 5/2x^2 + 5/4x^3 + 5/16x^4 + 1/32x^5$

(b) $\left(1 \; \underline{1} \atop \quad 20 \right)^5 = 1 + \underline{5}x\underline{1} \quad \underline{5} \; x \; \underline{1}$

$\quad\quad\quad\quad\quad\quad\quad\quad 2 \; 10 \; 2 \quad 100$

$1 \; \underline{11} \quad \text{or } 1.275$

$\quad 40$

296

7. (a) $a^6 - 6a^5b + 15a^4b^2 - 20a^3 b^3 + 15a^2 b^4 - 6ab^5 + b^6$

    (b) 60.256

8. $32 + 80x + 80x^2 + 40x^3 = 34.47$

9. (a) $1 + 5x + 10x^2 + 10x^3 + 5x^4 + x^5 = 0.8154$

    (b) 1.194

10. $1 - 15x + 90x^2 - 270x^3 = 0.8587$

11. (a) $1 + 5a + 10a^2 + 10a^3 + 5a^4 + a^5$

    (b) 0.9040

CHAPTER TWENTY EIGHT: PROBABILITY

1. (a)   p(both alive) = 0.7 x 0.9 = 0.63

    (b)   p (neither alive) = 0.3 x 0.1 = 0.03

    (c)   p (one live) = (0.7 x 0.1) + (0.9 x 0.3) = 0.34

    (d)   p (at least one alive)

          = (0.7 x 0.01) + (0.9 x 0.3) + (0.7x 0.9)

          = 0.7 + 0.27 + 0.63

          = 0.97

2. (a)   (i) P(B) = 8/15

          (ii) P (G or) = 7/15

    (b)   (i)   P (1st 2 pens picked are both green)

                = $^2/_{15}$ x ¼ = $^1/_{105}$ or $^2/_{210}$

          (ii) P (only one of the 1st 2 pens picked is red)

                = 8/15 x 5/14 + (2/15 x 5/14) + 5/15 x 8/14) + 5/15 x 2/14)

                = $\underline{40 + 10 + 40 + 10} = \underline{16}$

                   15 x 4   21

3. (a)   p(3 boys) = 1/22

    (b)   p (2 girls) =

          5/12 x 7/11 x 6/10 x 7/12 x 5/11 x 6/10 x 7/10 x 6/12 x 5/10

4.   (a)

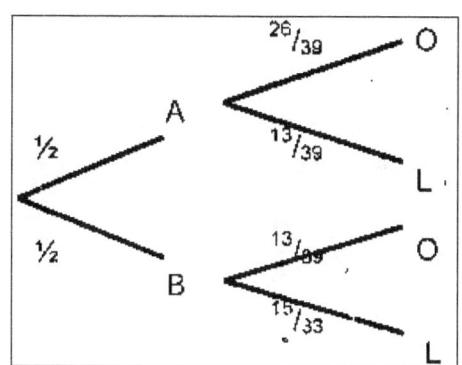

    (b)   p (orange) =(½ x 2/3) + (½ x 6/11)

             = 1/3 + 3/ 11

             = 20/33

5.   (a)   (i)   18/40 x 2/3 = 3/10

            (ii)   (18/40 x 2/3) + (22/40 x 3/5) = 63/100

    (b)   2/5 x 1/3 (18/40 x 22/39(+ 2/5 x 1/3 (22/40 x 18/ 39)

             = 22/325

6.   P (GGB) = 7/15 x 6/14 x 8/13

    P(GBG) = 7/15 x 8/14 x 6/13

    P (BGG) = 8/15 x 7/14 x 6/13

    P(2G + 1B) = (7/15 x 6/14 x 8/13) x 3)

     = 24/65 = 0.3692

7.   5/100 x 540 = 27

    80/100 x 180 = 144

    P(sick) = 171/720 = 19/80

     = 0.2375

8.   (a)

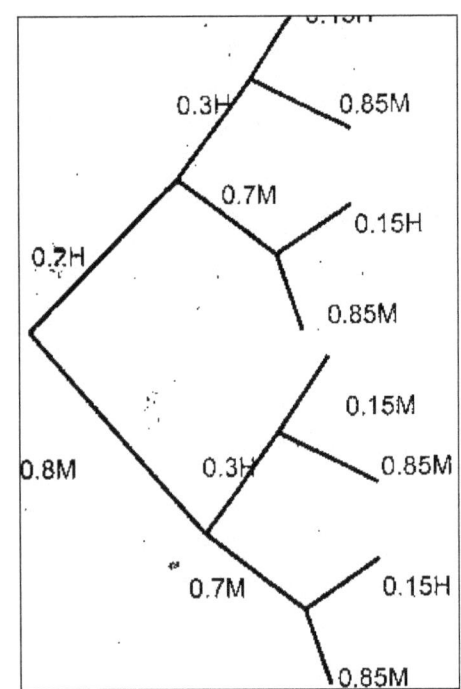

(b)   (i)   0.2 x 0.3 x 0.15 = 0.009

       (ii)   0.2 x 0.7 x 0.85 = 0.119

              0.8 x 0.3 x 0.85 = 0.204

              0.8 x 0.7 x 0.15 = 0.804

              0.407

       (iii)   HHM   0.2 x 0.3 x 0.85 = 0.051

              HMH   0.2 x 0.7 x 0.15 = 0.021

              MHH   0.8 x 0.3 x 0.15 = 0.036

              HHH   0.2 x 0.3 x 0.15 = 0.009

                                      0.117

9.   (a)   HHH, HHt, HTH, HTT

             TTT,TTh, THT, THH

(i) p (at least two heads) = 4/8 or ½

(ii) p (only one tail) = 3/8

(b) (i)

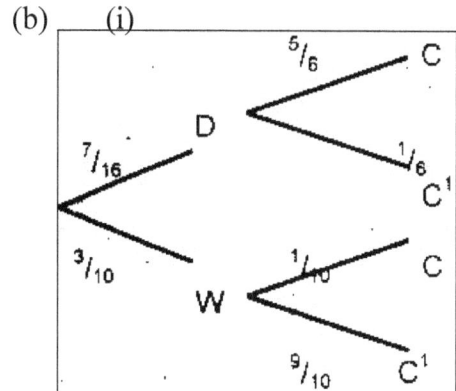

(ii) (7/10 x 5/6) + (3/10 x 1/10)

35/60 + 3/100 = 46/75

(iii) 3/10 x 9/10 = 27/100

10. Ratio 4:2:1

(a) (A wins) = 4/7

(b) P (either B or C wins)

= 2/7 + 1/7

= 3/7

11. 30/100 x 1.8 x 106 = 540,000

$$\frac{120,000}{1,000,000} \ x \ \frac{540,000}{1800000}$$

1/50 pr 0.02 pr 2%

12.

13. (a) P(RR) = 4 x 2 = 8

                    6   5   30

$P(YY) = {}^2/_6 \times {}^3/_5 = 6/30$

P (same colour) = 8/30 + 6/30

= 7/15

(b)   (i) $P(R_A R_A) = 4/6 \times 3/5 = 2/5$

                $P(R_B R_B) = 2/5 \times \frac{1}{4} = 1/10$ P (Both RED for A or B) = $2/5 + 1/10 = \frac{1}{2}$

        (ii)   P (all RED) = 2/5 x 1/10

                = 1/25

14.   (a) ${}^3/_{14}$

        (b) ${}^{41}/_{56}$

15.   (a) ${}^1/_{22}$

        (b) ${}^2/_{144}$

16. ${}^7/_{15}$

17. ${}^{24}/_{65}$

18. ${}^3/_{1024}$

19. ${}^{51}/_{60}$

20. ${}^{20}/_{33}$ or ${}^{260}/_{429}$ or ${}^{780}/_{1287}$

CHAPTER TWENTY NINE: COMPOUND PROPORTION AND MIXTURES

1. $\dfrac{(4 \times 21) + (3 \times 42)}{7} = 30$

    $\dfrac{130}{100} \times 30 = 39$

2. Cap of the tank

    $= 2.4 \times 2.8 \times 3 \times 1000$

    $= 20,160$ litres

    Amount needed

    $= 20,160 - 3,600$

    $= 16,560$ litres

    Time $= \dfrac{16560}{0.5 \times 60 \times 60}$ $\qquad = 9$ hrs 12 mins

3. (a) (i) Vol $= 135 \times 0.15 = 20.25 m^2$

        (ii) Mass $= 2500 \times 20.25$

            $= 50625$ kg (50630)

            $=$ mass of cement $= 50625 \times \frac{1}{9}$

            $= 5625$ kg (5625.56)

    (b) Bags of cement $= \dfrac{5625}{50}$

        $= 112.5$ or 113

    (c) No of lories of sand $50625 \times \underline{4}$

$$\frac{7000}{9}$$

= 3.214 = 4 lories

4. (a) Mass of maize in A 5/8 x 72 = 45 kg

    (b) Beans in A and B

    8/17 x 170 = 80 kg

    Maize in A and B

    9/17 x 170 = 90 kg

    Beans in B = 80 – 45

    = 53 kg

    Maize in B = 90 – 45

    = 45 kg

    Ratio 53.45   Or 1.1778.:1

5. (a) B.P per kg = $\frac{40 \times 65 + 60 \times 27.50}{100}$

    = Kshs 42.50

    (b) (i) S.P = $\frac{85 \times 120}{100}$

    Kshs 102 per packet

    (ii) New S.P = 102 x 90/ 100

    Kshs 91.80

(iii)   Total realized so far

(8 x 102) + (91.80 x 14)

= 816 + 1285. 20 = 2101. 20

Original total S.P = 102 x 50 = 5100

New price per packet

$= \dfrac{5100 - 2101 . 20}{28}$

$= \dfrac{2998.80}{28}$

Kshs 107.10

6.   Cost of beans in mixture = 3/5 x 2100

Cost of maize in mixture = 2/5 x 1200

Cost of mixture per bag = 3/5 x 2100 + 2/5 x 1200

7.   (a)   Volume = x. sec x length

= ½ x 25 (1 + 2.8) x 10

= 475 m^3

(b)   (i)   ½ x 25 x 1.8 x 10

= 225m^2

(ii)   Taken time to fill the tank

$\dfrac{9 \times 475}{225}$

= 19 hrs

∴ Time taken to fill remaining part

= 19 − 9

= 10 hrs

8.   (a)   Initial volume of alcohol

= 60/ 100 x 80 = 48 lts

New volume of solution = (80 + x)_ lts

$$\frac{48}{80 + x} = \frac{40}{100}$$

4800 = 3200 + 40x

40x = 1600

X = 40 lts

(b)   New volume of solution

= 80 + 40 + 30 = 150 lts

48/150 x 100 = 32

% age of alcohol = 32%

(c)   in lts

32% of 5 = 1.6 lts of alcohol

68% of 5 = 3.4 lts of water

In 2 lts 60% of 2   = 1.2 of alcohol

40% of 2   = 0.8 lts of water

In final solution (7lts)

2.9 lts are alcohol

4.2 lts are water

∴ Ratio of water to alcohol

$= 4.2 : 2.8 = 3 : 2$

Alternatively

(d)　　5 lts　W:A　　　$= 68:32$　　$= 17:8$

∴ Water　　$= 17/25 \times 5$　$= 17/5$

Alcohol　　$= 8/25 \times 5$　$= 8/5$

In 2 lts water　$= 40/100 \times 2$　$= 4/5$

Alcohol　　$= 60/100 \times 2$　$= 6/5$

Final solution

Water: Alcohol

$17/5 + 4/5 : 8/5 + 6/5$

$21/5 : 14/5$

$21 : 14 = 3 : 2$

9.　　(a)　　(i)　　Fraction filled in hr (P & Q)

　　$= 2/9 + 1/3 = 5/9$

　　Time taken to fill tank 1 4/5 hr

　　　　(ii)　　Fraction filled in 1 hr (P, Q & R)

$= 5/9 - \frac{1}{2} = \frac{1}{18}$

Time taken to fill tank $= 18$ hr

(b)　　(i)　　Fraction filled by 9.00 am

$P - \underline{2 \times 1h} = \underline{2}$

9   9

Q- $^1/_3$ x ¼ h = $^1/_{12}$

P & Q – $^2/_9$ + $^1/_{12}$ = 11/36

    (ii)   Fraction to be filled = $^{25}/_{36}$

Time tank will fill up 0900 + 1230

= 2130j (9.30 pm)

10.   2 $^{11}/_{12}$ hrs

11.   10 days

12. $^{3.5}/_{100}$ x 50 = 1.75

    (a)   (i)   Total = 3.175 kg

        (ii)   3. 969%

    (b)   A = 30kg

        B= 20 kg

        B ≥ 20kg

13.   3 ½ days

14.   (a)   OT = $^{12}/_7$p + $^3/_7$ r

        QT = $^3/_7$r - $^9/_7$p

        = 3/7 (r-3p)

    (b)   (i)   QR = r – 3p

            QT = $^3/_7$ QR

            ∴ QT & QR are parallel and Q is a common point

            ∴ Q, T and R lie on a straight line

        (ii)   QT : TR = 3:4

CHAPTER THIRTY: GRAPHICAL METHODS

1. (i) $b - a = 35$ ……. (i)

   $7b - 490a = 39$ ……(ii)

   $A = 4.9$ $b = 40$

   (ii) $S = 4.9t^2 + 40t + 10$

t	0	1	2	3	4	5	6	7	8	9	10
s	10		70.4	85.9	91.6	87.5	73.6		16.4	-26.4	

   (b) (i) Suitable scale

   Plotting

   Curve

   (ii) Tangent at t = 5

Velocity = -9.0 ± 0.5 m/s

   0.70 ± 0.1

2. (a) (i)

x	1.1	1.2	1.3	1.4	1.5	1.6
y	-0.3	0.5	1.4	2.5	3.8	5.2
X³	1.331	1.728	2.197	2.744	3.375	4.096

All values of x 3

All B1 for at least 4 or if all values are correct to 1 or 2 d.p

 (b) (i) Linear scale used

Line of best fit drawn 4 of this points correctly plotted

Plotting points

a=2

b = -3

(ii) $y = 2x^3 - 3$

3.   (a)   Log P = n log r + log K

(b)

P	1.2	1.5	2.0	2.5	3.5	4.5
Log P	0.08	0.18	0.30	0.40	0.54	0.64

R	1.58	2.25	3.39	4.74	7.86	11.5
Log r	0.20	0.35	0.53	0.68	0.90	1.06

Scale

Plotting

Line

Log k = 0.05

K = 2/3 = 0.6667

= 0.667 ± 0.0200

4.   Find midpoint (centre) = $\underline{5 + (-1)}$ $\underline{5 + (-3)}$

2   2

(a) $= (^4/_2, \, ^2/_2)$

$= (2, 1)$

(b)   Vector of   (a, b) = (2, 1)

$$R = \begin{pmatrix} 5 \\ 5 \end{pmatrix} - \begin{pmatrix} 2 \\ 1 \end{pmatrix} = \begin{pmatrix} 3 \\ 4 \end{pmatrix}$$

$$\therefore r = \sqrt{3^2 + 4^2}$$

$$= 5 \text{ units}$$

$$(x - 2)^2 + (y-1)^2 = 5^5$$

$$X^2 - 4x + 4 + y^2 - 2y + 1 = 25$$

$$X^2 + y^2 - 4x - 2y - 20 = 0$$

5. $n^2 - \frac{3}{2}x + (-\frac{3}{4})^2 + y^2 + y + (\frac{1}{2})^2 = -\frac{1}{4} + \frac{9}{16} + \frac{1}{4}$

$$= \frac{9}{16}$$

$$X - (\tfrac{3}{4})^2 + (y + \tfrac{1}{2})^2 = 9/16$$

Radius $= \frac{3}{4}$

Centre $(\frac{3}{4}, -\frac{1}{2})$

6.   (a)

Log x	-40	0.00	0.08	0.15	0.20
Log T	0.10	0.30	0.34	0.37	0.40

(b) (i) For all pts plotted

Apply ($\surd$) if at least B1 earned on table line of best fit drawn with at least 4 pts plotted.

(ii)   (a) $a = \log^{-1} 0.3 = 2.000$

           $B = \text{grad} = \underline{0.4 - 0.1}$ or equivalent

                $0.0 - (0.4)$

(c)   Log T = b log x + log a

Log x = $\underline{-0.3}$

0.5

X = 0.25

(d)   (ii)   Alternative

M log T = b x log a

Log T = b/m log x + 1/m log a

Intercept = 1/m log a = 0.3

=>a = log^{-1} 0.3 m

Grad = b/m = 0.4 – 0.1

0.1- (0.4)

B = 0.5 m

(e)   mlog T = n log x + log a

0 = 0.5m log x + 0.3m

Log x = $\underline{-0.3m}$

0.5m

X = 0.25

CHAPTER THIRTY ONE: MATRICES AND TRANSFORMATIONS

1.   (a) $\Delta = -3$

$$P^{-1} = \frac{1}{3} \begin{pmatrix} 8 & -7 \\ -5 & 4 \end{pmatrix}$$

(b)   (i)

$$\begin{pmatrix} 8 & 14 \\ 10 & 16 \end{pmatrix} \begin{pmatrix} b \\ m \end{pmatrix} = \begin{pmatrix} 47600 \\ 57400 \end{pmatrix}$$

(ii)

$$\begin{pmatrix} -8/3 & 7/3 \\ 5/3 & -4/3 \end{pmatrix} \begin{pmatrix} 8 & 14 \\ 10 & 16 \end{pmatrix} \begin{pmatrix} b \\ m \end{pmatrix} = \begin{pmatrix} -8/3 & 7/3 \\ 5/3 & -4/3 \end{pmatrix} \begin{pmatrix} 47600 \\ 57400 \end{pmatrix}$$

$2b \quad = \quad 7000$

$2m \quad \quad 2800$

Beans Kshs 3500

Maize 1400

(c)   New price of beans $= 105/100 \times 3500 \times 5$

Balance for maize $= 47600 - 29400$

$\quad\quad\quad = 18200$

Bags of maize $= \underline{18200} = 13$

New ratio = 8: 13

2.   A   B   C   A'   B'   C'

$$\begin{pmatrix} 0 & 1 \\ -1 & 0 \end{pmatrix} \begin{pmatrix} 2 & 4 & 1 \\ 1 & 4 & 6 \end{pmatrix} = \begin{pmatrix} 1 & 4 & 6 \\ -2 & -4 & -1 \end{pmatrix}$$

3.   (a) (i) Diagram

(ii) A" (1, 2) B (7, -2) C"(5, -4) D"(3, -4)

(b) A" (-1, 2) B" (-7, -2) C"(-5, -4) D"(-3, 4)

(c)   Half turn

       Centre (0,0)

4.   (a) (i) $\begin{pmatrix} a & b \\ c & d \end{pmatrix} \begin{pmatrix} 2 & 5 \\ 3 & 3 \end{pmatrix} = \begin{pmatrix} -4 & -1 \\ 3 & 3 \end{pmatrix}$

$2a + 3b = 4$ $2c + 3d = 3$

$5a + 3b = -1$ $5c + 3d = 3$

$A = 1, b = -2$ $c = 0, d = 1$

Therefore M = $\begin{pmatrix} 1 & -2 \\ 0 & 1 \end{pmatrix}$

(ii) $\begin{pmatrix} 1 & -2 \\ 0 & 1 \end{pmatrix}\begin{pmatrix} 4 & x \\ 1 & y \end{pmatrix}\begin{pmatrix} = 2 \\ 1 \end{pmatrix}\begin{pmatrix} \\ \end{pmatrix}$

$C_1 = 2,1$

(b) $\begin{pmatrix} 0 & 1 \\ 1 & 0 \end{pmatrix}\begin{pmatrix} \\ \end{pmatrix}\begin{pmatrix} 1 & -2 \\ 0 & 1 \end{pmatrix}\begin{pmatrix} = \\ \end{pmatrix}\begin{pmatrix} 0 & 1 \\ 1 & -2 \end{pmatrix}\begin{pmatrix} \\ \end{pmatrix}$

5.   (a) PR = $\begin{pmatrix} 0 & -1 \\ -1 & 0 \end{pmatrix}\begin{pmatrix} a & b \\ c & d \end{pmatrix} = \begin{pmatrix} -c & -d \\ -1 & -b \end{pmatrix}$

$\begin{pmatrix} -c & -d \\ -a & -b \end{pmatrix}\begin{pmatrix} 2 & 2 & 4 \\ 0 & 4 & 4 \end{pmatrix} = \begin{pmatrix} 0 & -4 & -4 \\ 2 & -10 & -12 \end{pmatrix}$

$-2c = 0 \qquad => c = 0$

$0 - 4d = -4 \quad => d = 1$

$-2a = -2 \qquad => a = 1$

$-2a - 4b = -10 => b = 2$

$\therefore R \begin{pmatrix} 1 & 2 \\ 0 & 1 \end{pmatrix}$

315

$$\begin{array}{ccc} \text{A} & \text{B} & \text{C} \end{array} \qquad \begin{array}{ccc} \text{A'} & \text{B} & \text{C} \end{array}$$

(b)$\begin{pmatrix} 1 & 2 \\ 0 & 1 \end{pmatrix}$ $\begin{pmatrix} 2 & 2 & 4 \\ 0 & 4 & 4 \end{pmatrix}$ $\begin{pmatrix} 2 & 10 & 12 \\ 0 & 4 & 4 \end{pmatrix}$

(c) A sheer transformation

X – axis invariant and $j(0, 1) \rightarrow j(2, 1)$

6.   (a)   (i) Graph

(ii) $\begin{pmatrix} 1 & 0 \\ 3 & 1 \end{pmatrix}$

(b)   (i) Graph

7.   (a) $A = \begin{pmatrix} {}^{15}/_{17} & {}^{8}/_{17} \\ {}^{8}/_{17} & {}^{15}/_{17} \end{pmatrix}$

(b) $\theta = 28^0\ 4'\ (28.07^0)$

(c) $({}^{-3}/_{17}, {}^{114}/_{17})$

CHAPTER THIRTY TWO: STATISTICS

1.   7.5 x 5/8 X 4

2.

Vel	19.5	39.5	59.5	79.5	99.5	119.5	139.5	159.5	179.5
Cf	9	28	50	68	81	92	97	99	100

(a)   Cumulative frequency

Linear scale

Plotting

Smoothing & complete of CF curve

(b)   (i)   Upper quartile = 90

Lower quartile = 36

Range = 90 – 36 = 54

(ii)   No. of days = 100 – 93 = 7

3.   25, 289, 4, 484, 4 806

$$J = \sqrt{\frac{806}{5}}$$

$$\sqrt{161.2}$$

= 12.7

4.

mdx	F	fx	Fx³
9	4	36	324
12	7	84	1008
15	11	165	2475
18	15	270	4860
21	8	168	3528
24	5	120	2880
Σ fx = 843			15075

Fx: 36, 84 165, 270, 168, 120

(a)   Mean = $\underline{843}$

                        50

= 16: 86

(b)   (i)   fx 2: 324, 1008, 2475, 4860, 3528, 2880

Variance = $\underline{15075 - 16.86}$

                        50

= 301.5 – 284.2

17.3 (17.24)

(ii)   S.D = $\sqrt{17}$: 3

            = 4.159 or (4.159 or (4.152)

5.

Class	14.5 – 18.5	18.5 – 22.5	22.5 – 26.5	26.5 – 30.5	30.5 – 34.5	34.5 – 38.5	38.5 – 42.5
Frequency	2	3	10	14	13	6	2
C. freq	2	5	15	29	42	48	50

Cumulative frequencies

(a)　　Linear scale used

　　　　Plotting of cf against upper class limit

　　　　Complete of cf curve drawn

(b)　(i)　　Median = 29.5

(ii)　Reading at mass $25 – 28 = 11$ and 20

　　　Probability = $\dfrac{20.\ 11}{50} = 0.8$

6.　　$\dfrac{3 \times 125 + 4 \times 164 + 2 \times 140}{3 + 4 + 2}$

　　= $\dfrac{1311}{9}$ ＝ $145^2/_3$

7.　　No of people = $\dfrac{360 \times 1080}{144}$

　　No of children = $2700 – (510 + 1080)$

　　　　　　　　　　　　　　= 1110

　　Angle of children　　$\dfrac{1110 \times 360}{2700}$

　　　　　　　　　　= 148^0

319

8.   (a)

X	1.0 – 1.9	2.0 – 2.9	5.0-3.9	1.0-1.9	5.0-5.9	6.0-6.9
F	6	11	10	7	2	1
d	6	20	30	37	39	40

Lower quartile = 1.95 + 1x 4/14 = 2.236 (2.24)

Upper quartile = 2.95 + 1 x 10/10 = 3.95

Inter quartile range = 3.95 – 2.236 = 1.714

(b)   x   f   dx –a   fd fd^2

       1.45   6   -2   -12   24

       2.45   14   -1   -14   14

       3.45   10   0   0   0

       4.45   7   1   7   7

       5.45   2   2   4   8

       6.45   1   3   3   9

                                -12   62

$$Sd = \sqrt{\frac{62}{40} - \frac{(-12)^2}{40}} = 1.55 - 0.09$$

$$\sqrt{1.46}$$

= 1.208

9.   (a)

320

Mass (g)	25 - 34	35 - 44	45 - 54	55 - 64	65 - 74	75 - 84	85 - 94
No. of potatoes	3	6	16	12	8	4	1
Cf	3	9	25	37	45	49	50
Upper class boundaries	34.5	44.5	54.5	64.5	74.5	84.5	94.5

(b) (i) Position of 60th percentile = $\dfrac{60 \times 50}{100}$

$\therefore$ Mass of 30th potato = 58.5g

60th percentile mass = 58.5g

(ii) No. of potatoes with mass of 53g or less = 28

No of potatoes with mass of 68g of less = 40

$\therefore$ No. of potatoes with mass of 53 to 68g = 40 − 28 = 12

$\therefore$ age of potatoes with mass 53g to 68g

$= \dfrac{12}{50} \times 100 = 24\%$

10. Area = A = 5 x 3.2

B = 10 x 1.2

16: 12 = f:6

12f = 96

F = 8

11. (a) (i)

Marks	0-10	10-30	30-60	60-70	70-100
Frequency					
Area of rect	60	200		40	120
Height of rect	6	10		4	4

(ii)

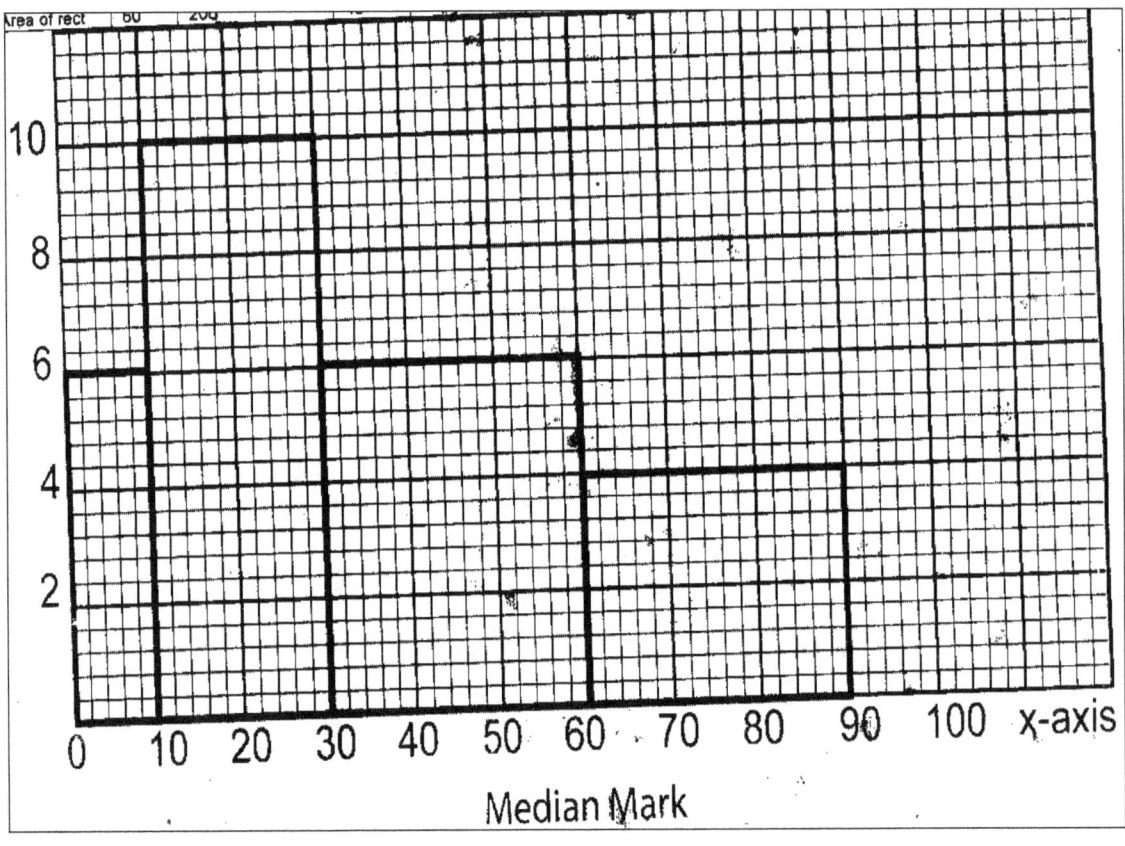

Histogram

(b)   (i) Median in group 30-60

(ii) $60 + 200 + 6x$

$= \frac{1}{2}(60 + 200 + 180 + 40 + 120)$

$260 + 6x = 300$

$X = 6 \frac{2}{3}$

$\therefore$ Median $= 30 + 6 \frac{2}{3}$

$= 36 \frac{2}{3}$

12.   (a)   3rd day $= 60$

               4th day $= 61$

(b) $M_3 = 61$

            $M_5 = 64$

13.   (a)   Ans 16, 8, 6

(b)   (i)   17.3

          (ii)   4.159

14. $M_1 = 50.99$

      $M_2 = 50.29$

      $M_3 = 50.65$

15.   (a)   Graph

        (b)   (i)   29.5

               (ii)   0.8

CHAPTER THIRTY THREE: LOCI

1. $<ABC = 105^0$ or $<BAD = 75^0$

Complete// gram constructed

Const. of loci: $AP \leq 4$ cm

$BQ \leq 6$ cm

Area// gram = $7 \times 10 \sin 105^0$

$= 7 \times 10 \times 0.9659$

$= 67.61$ cm^2

Total area of sectors

$$\frac{75}{360} \times \frac{22}{7} \times 4^2 + \frac{105}{360} \times \frac{22}{7} \times 6^2$$

$= 10.48 + 33 = 43.48$

Required area $= 67.61 - 43.48 = 24.13$

2.   (a)   Bisecting $< BAD$

(b)   Construction of 1 at B and at A construction of 45^0 or 135^0 to get $67 \frac{1}{2}^0$ at B construction of 1 Bisector of AB identification of AB identification of $\otimes$ the centre O. Identification of the locus P

(c)   Size of the $<ABC = 131 \pm 1^0$

3.   (a)   Construction of 30^0

Check for construction marks

(b)   CD = 5.4 cm or 5.4 ± 0.1

(c)   DA = 4.5 or AA' = 1.5

(d)   Line through parallel to BC

4.

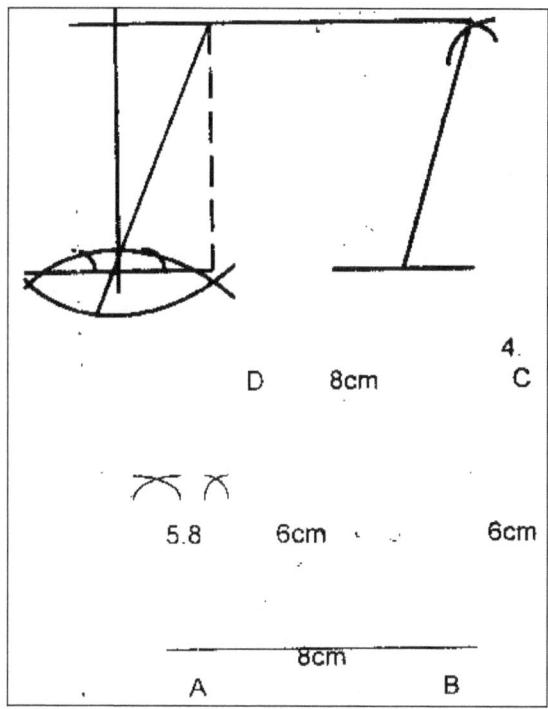

D   8cm   4. C

5.8   6cm   6cm

8cm

A   B

5.   (a)   Construction of 30^0

Completion of $\triangle$ PQR

(b) $\perp$ Bisector of PR (must be seen)

Location of S, QS = 8 cm and drawing $\triangle$ PRS

(c)   Construction of semi- circle with diameter SQ, Construction of parallel line to QS through R location of T_1 and T_2

6.

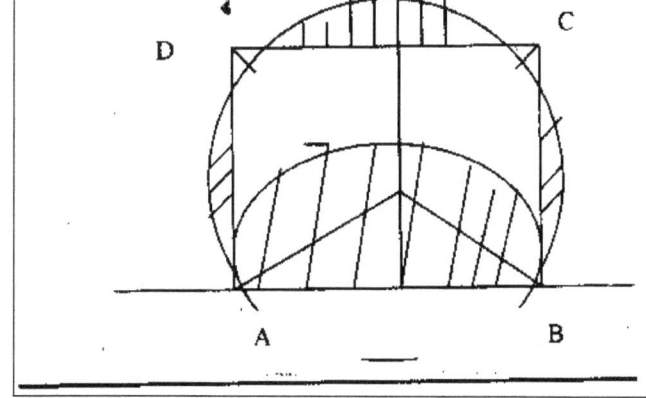

7.

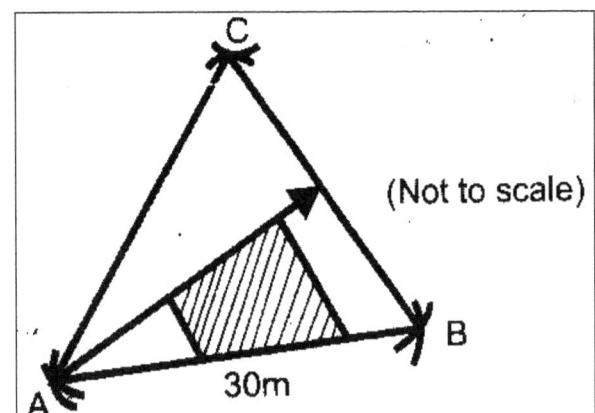

8.

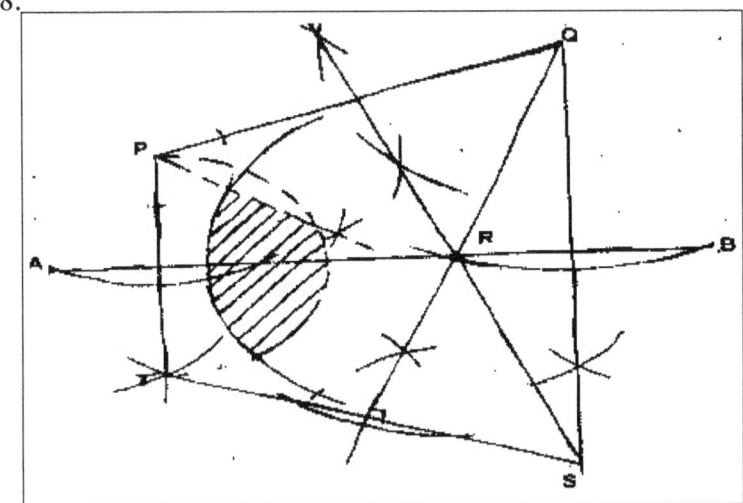

9.   (a)   Diagram

      (b)   (i) 73 ± 1 km

             (ii) $102^0 + 1^0$ or $578^0 E + 1^0$

10.

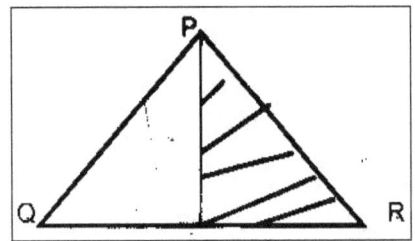

326

11. (a) Const of $\perp$ bisector of AB

(b) Const of $\perp$ bisector of AC or BC

$< OAB = 12^0 \perp 1^0$ or $< OBA = 12^0 \pm 1^0$

Position of P on XY and AB

CHAPTER THIRTY FOUR: TRIGONOMETRYII

1. (a)

X	0	10	20	30	40	50	60	70	80	90	100	110	120
Sin 3x	0	0.500	0.8660	1.000	0.866	0.500	0.000	0.500	-866	-100	-0.866	-0.500	0.000
2 sin 3x	0	1.00	1.73	2.00	1.73	1.00	0.000	-1.00	-73	-2.00	-1.73	-1.00	0.00

(b) Diagram on graph

    (i) Suitable linear scale

        Plotting

        Smooth sine curve

    (ii) $x = 76^0 \pm 1^0$

        $X = 104^0 \pm 1^0$

2. (a)

X	30	60	90	120	150	180	210	240	270	300	330	360
Cos X	0.87		0	-0.5		-1.0		-0.5	0	0.5	0.87	1.0
2 cos ½ X		1.73	1.41	1.0		0	0.52		1.41	1.7	1.93	

3.

	0	30	45	60	90	120	135	150	180	225	270	315	360
sin x	0	1	1.4	1.7	2	1.7	1.4	1	0	-1.4	-2	-1.4	0
os X	1	0.9	0.7	0.5	0	-0.5	-0.7	-0.9	-1	-0.7	0	0.7	1
	1	1.9	2.1	2.2	2	1.2	0.1	0.1	-1	-2.1	-2	-0.7	1

(b) Scale used

Plotting

Smooth curve

(c) $140^0 \pm 3^0 < 140 \pm 3^0$

4. (a)

X	0	10	20	30	40	50	60	70	
Tan X	0	0.8	0.36	0.58	0.84	1.19	1.73	2.75	5
2x + 30	30	50	70	90	10	130	150	170	
Sin (2x + 30)	0.50	0.77	0.94	1	0.94	0.77	0.50	0.17	(a)

X^0	0	30	60	90	120	150	180
$2 \sin x^0$	0	1	1.73	2	17.3	1	0
$1 \cos x^0$	0	0.13	0.5	1	1.5	1.87	0

(b) Graph

(c) (i) 126^0

(ii) $0^0 \le x \le 126^0$

329

6   (a)

X	30^0	105^0
Y	1.7	-2.4

(b)

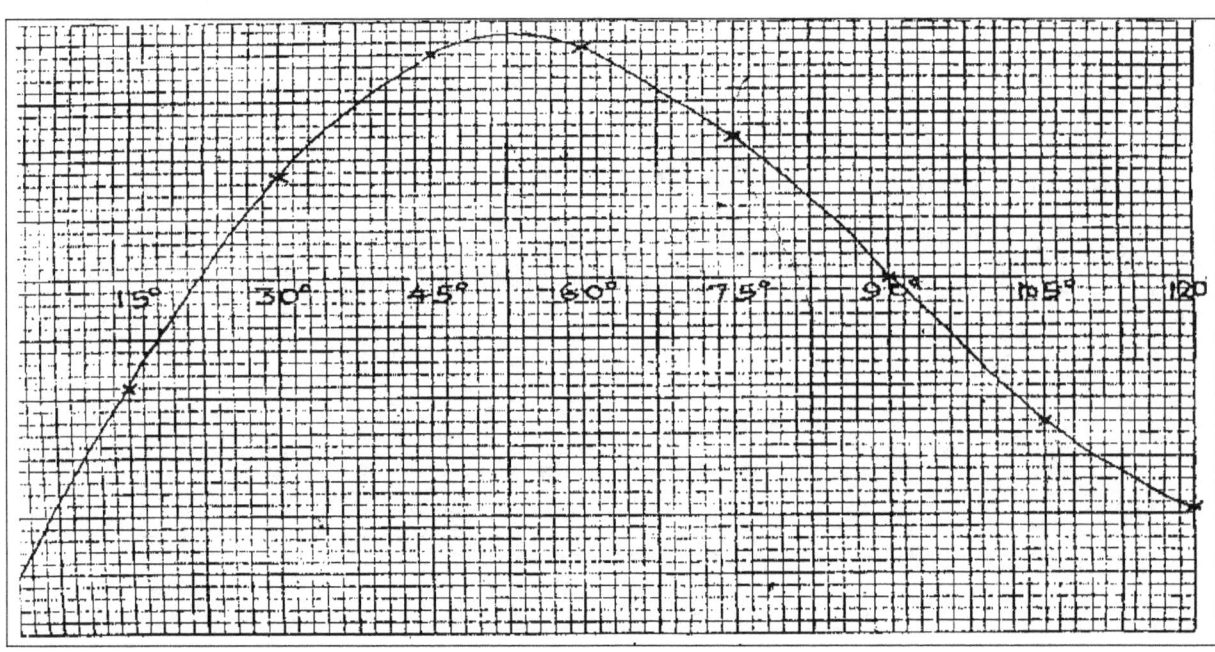

(c)   (i)   Maximum y = 4.1 ± 0.1

       (ii)   8sin 2x – 6 cos x = 2

              X = 31.5 ± 0.75^0

              X = 78 ± 0.75^0

7.   x = 0^0, 180^0, 360^0

8.   x = 0^0. 180^0

9.   131, 79^0, 228.21^0

10. $\underline{3 \tan θ}$ or 3 sec θ tan θ

       Cos θ

11.　　Sin d = -4/5 or – 0.8

　　　3rd quadrant 180 + 53.15 = 233.13^0

　　　4th quadrant 360 – 53.15 = 306. 87

12.　　Sin (90 x – x) = 8/10 = 4/5

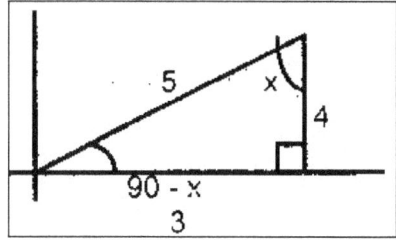

　　　Tan x = ¾

13.　　(a)

X	20	40	80	120	140	160	180
-3cos 2x^0	-2.30	-0.52	2.82	1.50	0.52	-2.30	-3.00
2b sin (3/2x^0 + 30^0	1.73	2	1.00	-1.00	-1	-200	-1.73

　　　(b)　　Roots x = 62 ± 2^0

　　　　　　X = 156 ± 2^0

14.　　131.79^0, 228.21^0

15. (a)

X	30	60	90	120	150	180	210	240	270	300	330	360
Cos X	0.87		0	-0.5		-1.0		-0.5	0	0.5	0.87	1.0
2 cos ½ X		1.73	1.41	1.0		0	0.52		1.41	1.73	1.93	1.0

(b)   Period = 720^0

Amplitude = 2

(c)   Enlargement of 2 about the centre

CHAPTER THIRTY FIVE: THREE DIMENSIONAL

GEOMETRY

1. (a) (i) $OA = (\sqrt{3^2 + 4^2})^{1/2}$

                $= 2.5$

                $VA = \sqrt{6^2 + 2.5^2}$

                $= \sqrt{42.25}$

                $= 6.5 \text{ cm}$

(ii)

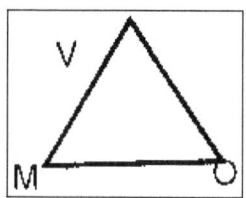

                        1.5

    (b) $\tan \beta = \dfrac{3}{2\frac{1}{4}} = 1.333$

        $\beta = 53^0 \, 7'$

        $\theta = 75^0 \, 58' - 53^0 7'$

        $= 22^0 \, 51'$

2.

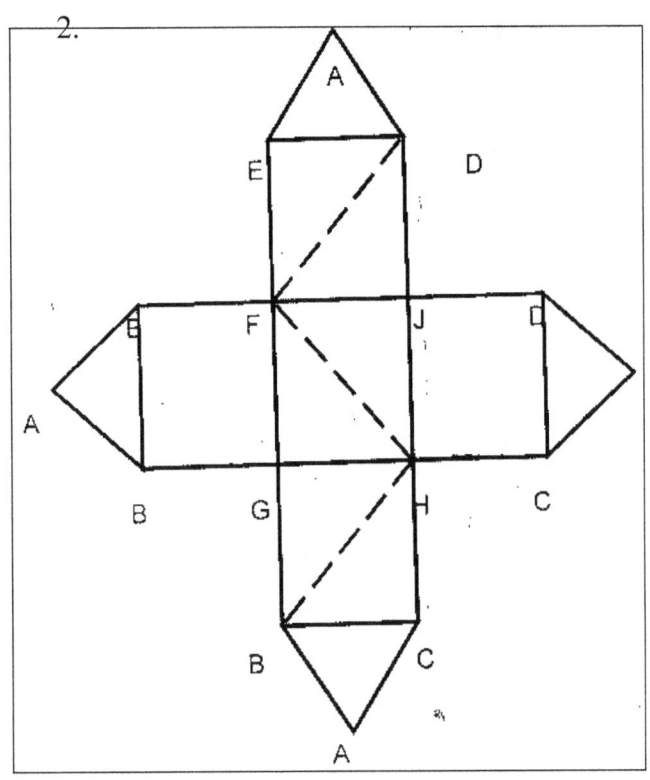

3.   (a) $FH^2 = 4.5^2 + 8^2 = 20.25 + 64$

$FH^2 = 84.25$

$FC^2 = 84.25 + 36 = 120.25$

$FC = \sqrt{120.25} = 10.97$ cm

(b)   (i)

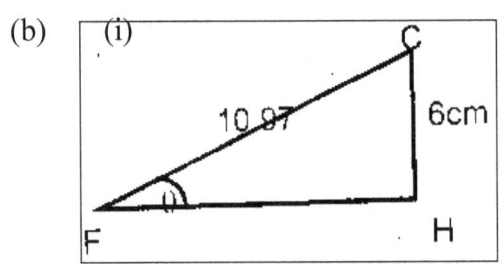

$\theta = 33.16^0$

(ii)   Tan θ = $\frac{4.5}{8}$ = 0.5625

θ = 29.36

(c)

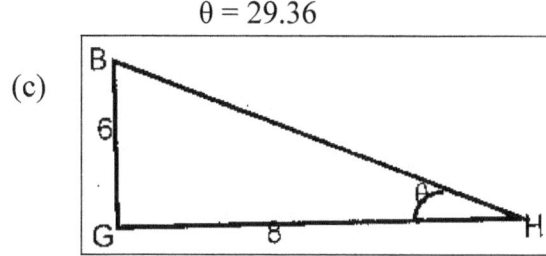

= 36.87^0

4.   (a)   Sketch

(b)   θ = 61^0 53 (61.88^0)

5.   (a)   (i)   BN = 8.65 cm

(ii)   EN = 13 cm

(b)   33^{0}40' (33.67^0)

CHAPTER THIRTY SIX: LATITUDES AND LONGITUDES

1.   (a)   (i)   Lat of B = 43.75^0 (43^0, 45')

            (ii)   r = 6370 cos 5375^0

                Angle between B and C = 60^0

           BC = $\dfrac{60}{360}$ x 2 x $\dfrac{22}{7}$ x 6370 cos 43.75

              = $\dfrac{60}{360}$ x 2 x $\dfrac{22}{7}$ x 6370 x 0.7224

              = 4820.816 km

        (b) $\dfrac{60 \times 4}{60}$ = 4 hrs

              Local time at C is 2100 hrs or 9.00 P.m

2.   (a)   Longitudinal difference = 70 – 10

        (b)   Distance between x and y

        (i) $\dfrac{60}{360}$ x $\dfrac{22}{7}$ x 2 x 6371 x cos 45

              $\dfrac{1}{6}$ x $\dfrac{22}{7}$ x 2 x 6371 x 0.7071

              = 4718 km

        (ii) $\dfrac{4919.45}{1.85}$ = 2551.05 nm

        (c)   Time difference = 60 x 4 = 240 min

                           = 4 hrs

              Local time at x = 6. p.m

3.   (a)   Angle change = 52 – 38.5

$$= 13.5^0$$

$$S = 2 \times 22/7 \times 6370 \times 13.5/ 360$$

$$= 1501.5 \text{ km}$$

(b) $\theta/_{360} \times 2 \times {}^{22}/_7 \times 6370 \cos 52^0$

$= 2400$

$$\theta = \underline{2400 \times 7 \times 360}$$

$$2 \times 22 \times 6370 \cos 52^0$$

$$= 35.05^0$$

$$C = (52^0 \ 21 \ W)$$

4.   (a) $60 \times 60 = 3600 \text{ nm}$

(b) $\theta = 31^0 \times 13^0 \text{ or } 18^0$

Distance from town A to town B

$= 60 \times 18 \cos 30$

$= 60 \times 18 \times 0. 8667$

$= 935. 28 \text{ nm}$

Total distance $= 935.28 + 3600$

$= 4535.28 \text{ nm}$

Total time $= \underline{4535. 28} + 0.25$

200

$22.6764 + 0.25$

22.926 h

Or 22 h 55.6 min

5.   (a)   Difference in time = 3 hrs

$\therefore$ Longitude difference = 3 x 15^0 = 45^0

Longitude of B = 15^0 + 45^0 = 60^0E

(b)   (i)   Distance traveled = 850 x 3 ½ km

= 2975 km

Arc AB = 2975

45/360 x 3142 x 2r = 2975

R = 3788 or 3787 or 3789

(ii)   6371 cos θ = 3788

Cos θ = $\dfrac{3788}{6371}$ = 0.594

θ = 53.51

Latitude of the two towns is 53. 51^0N

6.   Longitude difference = 360 – (133^0+ 118^0)= 109^0

109^0 x 60 cos x = 5422

Cos x = 0.8291

X = 33.99^0

Latitude of A and B is 34^0 N

7.   (a)   = 13 347 km

(b)   16.68 hrs

8.   (i)   = 7200 nm

(ii)   = 9353 nm

9.   (a)   250 km

(b) $137^0\ 27$'

CHAPTER THIRTY SEVEN: LINEAR PROGRAMMING

1.   (a) $x \geq 0$ and $y \geq 0$

               $X + Y \geq 7$

               $64x + 48y \geq 384$ $(4x + 3y \geq 24)$

   (b) $x + y = 7$ drawn

               $64x + 48y = 484$ drawn

               Shading

   (c)   No. of buses for minimum cost 3 type x and 4 type y or for x = 3 and y = 4

2.   (a) $x + y \leq 500$

               $Y > x$

               $X \geq 200$

   (b) $x + y \leq 500$ drawn and shaded

               $Y > x$

               $X \geq 200$

   (c)   (i)   No enrolled in technical = 249

                       No enrolled in business = 251

            (ii)   Max profit

                       249 x 2500 + 251 x 10000

                       = 873500

3.   (a) $x + y \leq 400$, $x > y$, $x \leq 300$, $y \geq 80$

   (b)   All 4 inequalities $\sqrt{}$ y drawn and shaded.

   (c)   (i)   x = 300 and y = 100

            (ii)   Max profit = 600 x 300 + 400 x 100

$$= 220,000$$

4. (a) $x \geq 0, y \geq 0, x + y \leq 6$

$$25x + 50y \geq 175$$

$$30x + 45y \geq 180$$

(b) $x \geq 0$
      $X + y \leq 6$

$25x + 50y \geq 175$ Correctly drawn and shaded

$30x + 45y \geq 180$

(c) Minimum cost at $x = 5$ and $y = 1$

Minimum cost $= 5 \times 20 + 1 \times 50 = 150$

5. (a) $300x + 180y \leq 1800$

$5x + 3y \leq 300$

$X + y \leq 80$

$X > 0, y > 0$

(b) $5x + 3y \leq 300$
    $X + y \leq 80$ Correctly drawn and shaded

(c) $x = 30 \; y = 50$

Maximum profit in Kshs $= 50 \times 4000 + 30 \times 6000$

$= 380, 000$

6. $2x + 5y \leq 40$

$5x + 8y \leq 80$

$X \geq 3$

$Y > 1/3\ x$

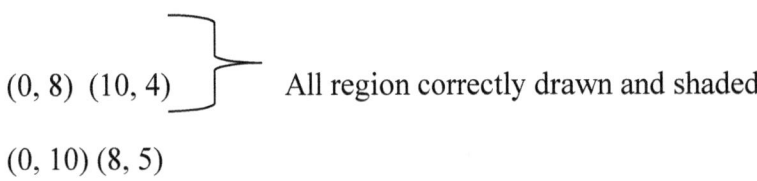

$(0, 8)\ (10, 4)$ All region correctly drawn and shaded

$(0, 10)\ (8, 5)$

Search line with gradient -3/5 drawn

Type A = 9

Type B = 4

CHAPTER THIRTY EIGHT: CALCULUS

1. Area = 2(8 + 6.5 +5.6+6+6.4 + 4.7(x 25

    = 2 x 37.2 x 25 x 100 or equivalent

    186000 ha

2. Choose positive roots only

    Integrate

    Substitute numerals

    Ans = 110. 38

    Or

    108 + 2 = 110

3. Missing values of y; 26, 138

    Area = ½ x 2 (10 + 230) + 2(6+26+70+138)

    = 240 + 480

    = 720

4. 3.55 ± 0.05, 4.85 ± 0.05, 5.7, 6.3, 6.7 & 6.9

    Area = ½ x 1 {0 +7 + 2 (3.6 + 4.9 + 5.7 + 6.3 + 6.7 + 6.9)}

    = ½ x 1 {7 ± 68.2)

5. (a) $x^2 - 2x - 3 = 0 <=> (x-3) (x+ 1) = 0$

    X = 3 or x -1

    (b) $(x^2 - 2x - 3) dx = x^3/3 - x^2 - 3x + c$

    (c) $\left[x^3/3 - x^2 - 3x \right]^{3}/_{2} = \left(27/3 - 9 - 9 \right) - \left(8/3 - 4 - 6 \right)$

    = 1 ²/₃

$$\left[X^3/3 - x^2 - 3x\right]^4 \quad = \quad \left[^{64}/_3 - 16 - 12\right] - \quad \left[^{-27}/_3 - 9 - 9\right]$$

$2\,^1/_3$

Sum of arcs = -1 2/3 + 2 1/3

$= 1^2/_3 + 2\,^1/_3$

= 4

6. (a)

X	2	3	4	5	6	7	8
Y	3	5	9	15	23	33	45

(b) A = ½ x 1 x {(3 + 45) + 2 (5 + 9 + 15 + 23 + 33)}

= ½ (48 + 170

= 109 (109. 25)

(c) -8

$\int (x2 - 3x + 5)dx$

2

$= \dfrac{x^3}{3} - \dfrac{3x^2}{2} + 5x$

$\dfrac{8^3}{3} - \dfrac{3 \times 8^2}{2} + 5 \times 8) - \dfrac{2^3}{3} - \dfrac{3 \times 2^2}{2} + 5 \times 2$

= 108

344

(d)   It would give an underestimate because the lines for the trapezia run below the curve in the region

7. $(x^2 + 1)(x - 2) = x^3 - 2x^2 + x - 2$

$\dfrac{dy}{dx} = 3x^2 - 4x + 1$

When $x = 2$ $\dfrac{dy}{Dx} = 5$

$Y = 0$

$\dfrac{y - 0 = 5}{x - 2}$

$y = 5x - 10$

8.   (a) $V = \dfrac{ds}{dt} = 3t2 - 5t + 2$

$a = \dfrac{dv}{dt} = 6t - 5$

(b) $6t - 5 = 0$

$T = 5/6$

$V = 3\left(^5/_6\right)^2 - 5\left(^5/_6\right) + 2$

$= ^{25}/1_2 - ^{25}/_6 + 2$

$= ^{-1}/_{12}\,(0.0833)$

9.   (a) $\int (2x + 3 x 2)\,dx = x2 + x3 + c$

(b)   Area below x – axis

$[X^2 + x^3] = 0 - [(-2/3)^2 + (-2/3)^3]$

$= 0 - (4/9 - 8/27)$

$= 4/27$

Area above x – axis

$[x^2 + x^3] = [4 + 8] - 0 = 12$

Total Area $= ^4/_{27} + 12$

$= 12\ ^4/_{27}$

10.   Distance = 5/12 {2.6 + 2(2.1 + 5.3 + 5.1 + 6.8 + 6.7+4.7)}

$= 5/2\ (2.6 + 6.14)$

$= 160m$

11.   (a) $\int(2x^2 - 5)\ dx = 2/3\ x^3 - 5x + c$

Y $= 2/3x^3 - 5x + c$

$3 = 2/3 \times 8 - 5 \times 8 + c$

C $= 7\ 2/3$ OR $23/3$

Y $= 2/3 \times 3 - 5x + 7\ 2/3$

(b) $\int(2t^3 + t^2 - 1)\ dt = ^2/_4\ t^4 + m^1/_3\ ^3 - t + c$

$(^2/_4\ t^4 + ^1/_3{}^3 - t + c)^3 = (^2/_4 \times 3^4 + ^3/_3{}^3 - 3) - (^2/_4 + ^1/_3 - 1)$

$= (8\ ½ + 9 - 3) - (½ + ^1/_3 - 1)$

$= 46\ ½ - (^{-1}/_6)$

$= 46\ ^2/_3$

12. (i) $\underline{dy} = 6x^2 + x + -4$

    dx

    When x = 1

    $\underline{Dy} = 6 + 1 - 1$

    Dx

    = 3

    (ii) $y + \frac{1}{2} = 3(x-1)$

    $Y = 3x - 3 - \frac{1}{2}$

    $Y = 3x - 3\frac{1}{2}$

13. (a) Gradient = -1

    $Y = -x + 7$

    (b) $7 - x = (x-1)^2 + 4$

    $X^2 - x - 2 = 0$

    $(x-2)(x+1) = 0$

    X = 2, x = -1

    X = 2 when y = 5

    X = -1 when y = 8

    Coordinates of P, Q are P (-1, 8), Q (2, 5)

14. (a) $a = 25 - at^2$

    $V = \int(a)\, dt$

    $= \int(25 - at^2)\, dt$

$$= 25t - at^3/3 + c$$

$$V = 25t - 3t^3 + C$$

When $t = 0$ $V = 4ms-1$

$\therefore C = 4$

$$V = 25t - 3t^3 + 4$$

(b)   When $t = 2$

$$V = 25 \times 2 - 3 \times 8 + 4$$

$$= 50 - 24 + 4$$

$$= 30m/s$$

15.   (a) $\underline{dy} = x^2 + 2x - 3$

dx

(b) $x^2 + 2x - 3 = 0$

$(x + 3)(x-1) = 0$

$X = -3$ or $x = 1$

When $x = -3$

$Y = 11$

When x -1

$Y = 1/3$

16. (a)

X	0	0.4	0.8	1.2	1.6	2.0
Y	2.00	1.96	1.83	1.60	1.20	0

17. (a)　　$S = 5^3 - 5(5^2) + 3(5) + 4$

　　　　$= 125 - 125 + 15 + 4$

　　　　$= 19m$

(b)　　$V = ds$

　　　　　Dt

　　　$= 3t^2 - 10t + 3$

　　　$= 3(5)^2 - 10(5) + 3$

　　　$= 75 - 50 + 3$

　　　$= 28ms^{-1}$

(c)　　At rest $V = 0$

　　　$\therefore 3t2 - 10t + 3 = 0$

　　　$(3t - 1)(t - 3) = 0$

　　　$T = 1/3$ seconds or $t = 3$ seconds

(d)　　$a = dv$

　　　　　Dt

　　　$= 6t - 10$

　　　$= 6(2) - 10$

　　　$= 2\ ms^{-2}$

18　(a)　　$P(1, 3), (4, -12)$

(b)　(i)　　102/3 sq units

349

(ii)   13 1/3 Sq. units

19. $\dfrac{dy}{dx} = 3ax^2 + b$

3a + b = -5

A + b = 1

A = -3

B = 4

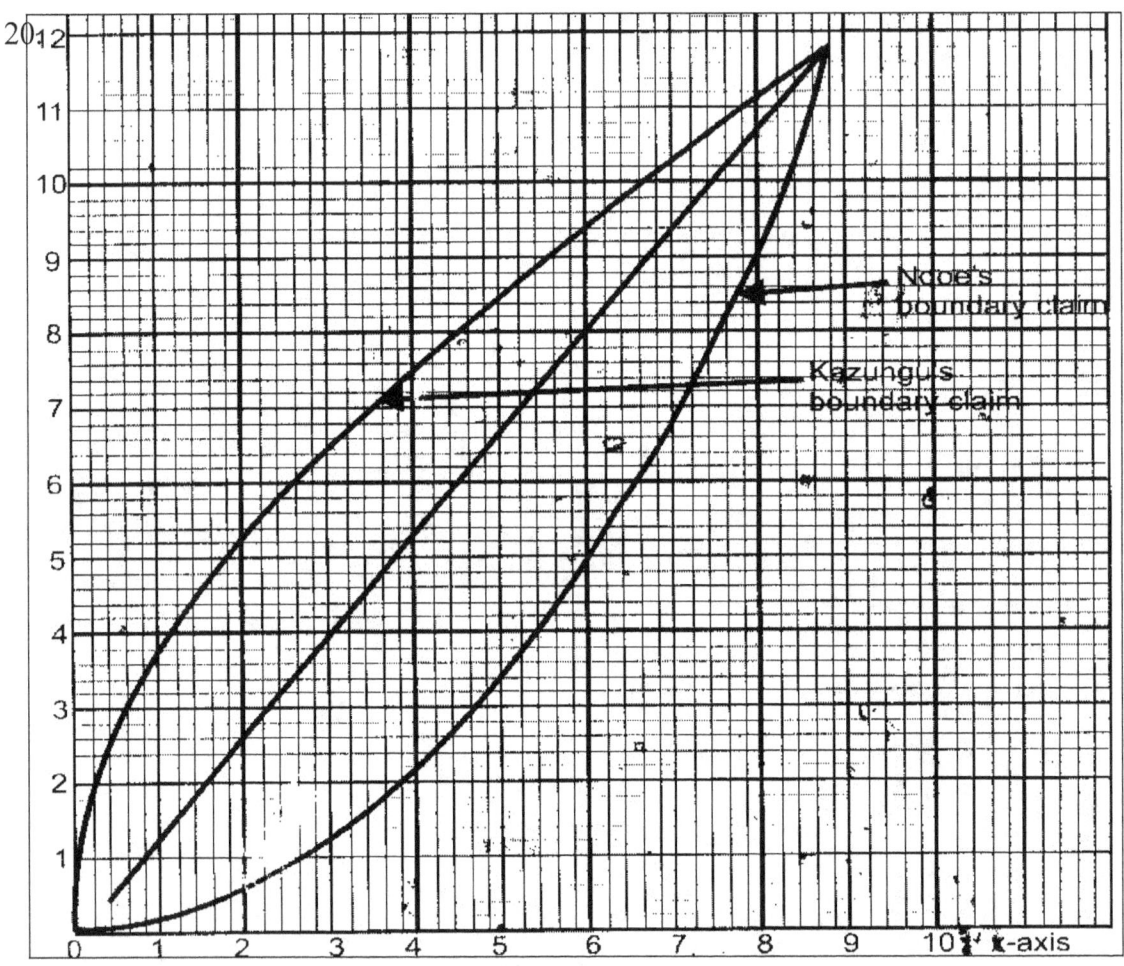

(a)   Curve y_1 y drawn

Curve y_2 y drawn

(b) (i) Area below upper curve

$$\frac{1}{2} \times 1 \times \quad 12 + 2 (4 + 5.7 + 6.9 + 8 + 9 + 9.8 + 10.6 + 11.3$$

$$\frac{1}{2} (12 + 130.6) = 71.3$$

Area below lower curve

$$\frac{1}{2} \times 1 \quad 12 + 2 (0.2 + 0.6 + 1.3 +)$$
$$24 + 3.7 + 5.3 + 7.3 + 9.5$$

$$= \tfrac{1}{2} (12 + 60.6) = 36.3$$

Area in dispute = 71.3 – 36.3 = 35

(ii) Area in hectares = $\dfrac{35 \times 400}{10000}$ = 1.4

21. (a) (i) $y = \dfrac{2x^2}{2} + x + c$

At x = -4, y = 6

$6 = (-4)^2 - 4 + c$

C = -6

$Y = x^2 + x - 6$

(ii) $x^2 + x - 6 = 0$

$(x-2)(x+3) = 0$

$X = 2$ or $x = 3$

$$\int_{-3}^{2} (x^2 + x - 6)\, dx = \left(\frac{x^3}{3} + \frac{x^2}{2} - 6x \right)_{-3}^{2}$$

$$= \left(\frac{8}{3} + \frac{4}{2} - 12 \right) - \left(\frac{-27}{3} + \frac{9}{2} + 18 \right)$$

$-7\frac{1}{3} - 13\frac{1}{2} = -20\frac{5}{6}$

Area $= 20\frac{5}{6}$

22. $S = 2t - \frac{t^2}{2} + c$

When $S = s$, $t = 2$

$\therefore 5 = 2 \times 2 - \frac{2^2}{2} + c$

$C = 3$

Thus $s = 2t\ \frac{1}{2}\ t^2 + 3$

23.   110.sq unit

24.   Missing values of y 26, 138

Area $= 720$ sq units

352

25. (a) x = 3 or -1

(b) $\dfrac{x^3 - x^2 - 3x + C}{3}$

(c) 4 sq. units

26. y = 5x – 10

27. (a) a = 6t – 5

(b) -1/12 m/s

28. (a) $x^2 + x^3 + C$

(b) 124/27 sq units

29. (a) Gradient = 4

(b) y = 4x – 1

30. (a) 4m/s

(b) (i) 4 22/27

(ii) 4 m/s^2

31. (a) 3m/s^2

(b) (i) t = 1 second or ½ second

(ii) $S = -1 \, {}^7/_8 \, m$

www.ingramcontent.com/pod-product-compliance
Lightning Source LLC
Chambersburg PA
CBHW080757180526
45168CB00006B/2244